Flying Machines Over Pensacola

AN EARLY AVIATION HISTORY FROM 1909 TO 1929

Commander Leo F. Murphy, U.S. Navy

Published by Pensacola Bay Flying Machines Ltd Co

PO Box 7176

Daytona Beach, Florida 32116

Cover

The photograph is of the 1909 Rhodes Airship, the earliest known aeroplane in Pensacola.

The topmost flying boat is a draftsman's drawing of a 24 passenger all-metal flying boat to be built by the Pensacola Metal Aircraft Corporation in the early 1930s.

ISBN: 0-9743487-0-8

Library of Congress Control Number: 200309558

First Edition: December 2003
(Centennial of Powered Flight)

Second Printing 2013

Contents

Contents

INTRODUCTION

The 20 years documented in this book may seem somewhat inadequate when one considers Pensacola's nearly 90 years of continuous aviation history. So many exciting aviation adventures and misadventures are compressed within these two decades, however, that to look further would compromise our appreciation for the challenges facing our early Gulf Coast aviators. In no other era in the development of flight would advances in the understanding of aeronautics come so quickly, the boundaries of aircraft performance be pushed so extremely, or the achievements of aviators be dramatized so publicly. For Pensacola, these

The caption from this November 9, 1910 cartoon reads "The way our necks will grow in time if this interest in aviation continues." (Courtesy of The Pensacola Journal)

two decades span the difference between the wild take-off of a homemade biplane built by two U.S. Army officers at Fort Barrancas in 1909 to the unruffled alighting of a twin-engine 22-passenger seaplane from the NYRBA airline upon Pensacola Bay in 1929.

Pensacola's rich aviation history truly begins, however, with the 1914 arrival of an antiquated battleship and sea-worn naval collier at the abandoned Pensacola Navy Yard. Aboard these ships were the entire fledgling Naval Flying Corps, aeroplanes and men, and from the day they stepped ashore Pensacola and Naval aviation would be forever synonymous. Think of the aeronautical marvels everyday Pensacolians delighted in courtesy of the nearby Naval Aeronautic Station! While most of the world had never seen an aeroplane, Pensacolians enjoyed an exclusive front-row seat to the most thrilling flying shows imaginable. Aviation progress in Pensacola was made not just in skips and hops but in incredible leaps and bounds as the newest, fastest and biggest aeroplanes and airships were brought to the station's expert pilots for evaluation and experimentation. Further, the U.S. Navy not only trained the first aviators in Pensacola, but also had the courtesy to outfit these early aviation entrepreneurs with surplus flying boats when they left the service, providing for the first tentative commercial aviation ventures in Pensacola.

The 1920s also presented several unique aviation challenges to Pensacola. Among these was the very real possibility that all Naval aviation training activities might be transferred as landplanes replaced seaplanes in the U.S. Navy. Additionally, as airmail and airlines expanded throughout the United States, Pensacola needed to find a way to attract these services here. In each case, Pensacola's civic leaders proved to be farsighted, innovative and often aggressive in their solutions.

Yet after scouring local historical resources, my research has revealed that many details on Pensacola's early aviation history, both military and civilian, remain to be discovered and documented. My hope is that this small effort will stimulate interest in this era and be but the first of many revised editions of this book as additional information is obtained. I welcome your contributions as we seek together to further our knowledge of Pensacola's early aviation history.

Where possible, I have let the pilots, observers, civilians and reporters tell their stories in their own-recorded words. I have found them to be more eloquent than I could ever hope to be. My sincerest thank you goes to those many unknown reporters of The Pensacola Journal who crafted their trade so well, and to the Pensacola News Journal for letting me publish them.

Commander Leo F. Murphy, United States Navy

Pensacola, Florida

ACKNOWLEDGEMENTS

SCENE IN PENSACOLA BAY, PENSACOLA, FLA.

(Courtesy Pensacola Historical Society)

As no one had ever captured Pensacola's early aviation history before in such detail, this book required untold hours of personal research in the various libraries, historical societies and museums of Pensacola. Accordingly, any errors, mistakes or omissions are the sole responsibility of the author.

During my explorations I was extremely fortunate to meet some very special people who provided me gentle guidance, invaluable assistance, and many times unlimited access to their priceless holdings.

In particular, I would like to thank Mr. Dean DeBolt and Ms. Katrina King of the Special Collections Department of the University of West Florida; Mr. Hill Goodspeed of the National Museum of Naval Aviation; Ms. Judith A. Walker and Ms. Jennifer Berg of the Naval Air Station Pensacola Library; Ms. Lynne Robertson and Mr. Lowell Bassett of West Florida Historic Preservation Inc.; Ms. Sandra Johnson and Mr. Dan Scott of the Pensacola Historical Society; Ms. Ann Hassinger of the United States Naval Institute; Commander Doug Seigfried, USN (Ret) of the Tailhook Association; Ms. Deborah Dunlap of Historic Pensacola; and Mr. Warren Brown of the Florida Aviation Historical Society.

I am deeply indebted to Ms. Lani Suchcicki of the Pensacola News Journal, for graciously obtaining permission for me to quote passages and use photographs from The Pensacola Gazette, The Pensacola Journal, Pensacola Evening News and Pensacola News Journal.

Lastly, I dedicate this book to my beautiful wife Denise and my wonderful children Kelli, Ben, Joel and Colin. Never has a man been so blessed by the love, patience and companionship of a family that has willingly and enthusiastically followed him across the oceans of the world in the service of his country.

Flying Machines Over Pensacola
A Selected Timeline

December 17, 1903	Wright brothers make the first powered, controlled aeroplane flight in history at Kitty Hawk, North Carolina.
November 14, 1909	Astonishing front-page headline in *The Pensacola Journal* announces that Lieutenant Rhodes of Fort Barrancas quietly has been building an aeroplane of his own design.
January 25, 1910	Lieutenant Rhodes tows his aeroplane behind a car at Fort Barrancas for his first unpowered flight. He later crashes shortly after take off in his first powered flight attempt.
February 1910	Lincoln Beachey makes the first heavier-than-air flight in Florida's history in Orlando.
November 8, 1910	Anthony Nassr makes the first lighter-than-air flight over Pensacola.
June 13, 1911	Anthony Nassr's attempt to make the first heavier-than-air flight in Pensacola's history ends in a crash when he collides with a tree stump on his take-off run.
October 20, 1911	Pensacola Navy Yard closed.
November 30, 1911	Nels J. Nelson makes the first successful heavier-than-air flight in Pensacola's history.
October 7, 1913	A special board is convened to prepare a comprehensive plan for Naval aeronautics. One of their recommendations is to concentrate all Naval aviation activities at Pensacola.
December 5, 1913	U.S. Marines reopen Pensacola Navy Yard as an advance base.
January 20, 1914	U.S. Navy aviation training unit arrives at the Pensacola Navy Yard to establish a Naval Aeronautical Station and Naval Flying School.
February 2, 1914	Lieutenant John Towers and Ensign Godfrey Chevalier make the first Naval aviation flight at Pensacola.
February 16, 1914	First aviation fatality in Florida's history occurs when Lieutenant (Junior Grade) Murray drowns in Pensacola Bay after crashing in a Burgess-Dunne seaplane.
March 3, 1914	Dr. M.E. Quinn becomes the first civilian Pensacolian to fly when special permission is obtained from the U.S. Navy for a flight in a Naval aeroplane.
April 20, 1914	First of two aviation detachments from Pensacola are sent to Mexico to fly scouting and reconnaissance missions in support of the Mexican War.

August 7, 1914	Ensign Chevalier makes the first night flight in Pensacola's history. Plane bursts into flames but lands safely before it is engulfed.
April 16, 1915	Lieutenant (Junior Grade) Bellinger makes a successful catapult launch from a stationary coal barge in Pensacola Bay.
June 1915	U.S. Navy orders a kite balloon, free balloon and non-rigid airship, all to be shipped to Pensacola.
July 4, 1915	Lieutenant (Junior Grade) Bellinger makes the first flight over Milton.
November 5, 1915	Lieutenant Commander Mustin makes the first catapult of an aeroplane from a ship when he launches from the USS NORTH CAROLINA making sternway in Pensacola Bay.
November 26, 1915	Wright Aeroplane Company visits Pensacola to look for flight training site.
April 2, 1916	Residents of Woolsey are officially notified to move or demolish their homes so the land could be cleared for a balloon field.
April 3, 1916	First flight of a kite balloon at NAS Pensacola.
May 5, 1916	First flight of a free balloon at NAS Pensacola.
July 12, 1916	Lieutenant Chevalier makes the first catapult from a ship underway when he launches from the USS NORTH CAROLINA in Pensacola Bay.
February 13, 1917	At Pensacola, Captain Evans, USMC, makes the first loop of a N-9 seaplane in history and recovers from a spin, he former thought impossible and the later unprecedented in the United States.
April 20, 1917	First flight of the U.S. Navy's first non-rigid airship, DN-1, at NAS Pensacola.
Apr 1917–Nov 1918	United States participation in World War I.
June 5, 1917	First American forces to land in Europe after war is declared are the First Aeronautic Detachment, comprised mostly of Naval aviators from Pensacola.
August 29, 1917	First flight of the B-class airship at NAS Pensacola.
December 7, 1917	Pensacola redesignated from Naval Aeronautical Center to Naval Air Station.
January 17, 1919	U.S. Naval Airship School established at NAS Pensacola.
February 4, 1919	Two U.S. Army Air Service landplanes conduct the first aerial voyage between Tallahassee and Pensacola.

April 28, 1919	Carried as a passenger, Yeomanette Thema DeBroux makes the first flight of a woman over Pensacola. There are two claims for the first woman to pilot an aeroplane over the city.
October 21, 1919	Mr. J.A. Whitted, a former Naval aviator piloting a surplus U.S. Navy flying boat, establishes the first commercial seaplane service at Pensacola.
June 25, 1920	Kiwanis Field, operated by civilians and located north of Pensacola, opens as the first landing field in Pensacola.
August 1921	Lighter-than-air training discontinued at NAS Pensacola.
September 21, 1921	Joint committee of the Chamber of Commerce, Kiwanis and Rotary Club offers Kiwanis Field to the U.S. Navy as a landing field.
December 1921	U.S. Navy accepts Kiwanis Field to establish landplane flight training in addition to the current seaplane flight training program.
February 21, 1922	Work begins at NAS Pensacola to enlarge and improve the balloon field for landplane use.
July 1, 1922	Landplane training commences at NAS Pensacola.
December 7, 1922	Kiwanis Field is dedicated and renamed Corry Field in honor of Florida native Lieutenant Commander William Corry, Naval Aviator Number 23, who died in the line of duty.
December 14, 1922	First landplane crash at NAS Pensacola.
March 26, 1923	Gulf Coast Air Line inaugurates seaplane service between Pensacola and Mobile.
April 17, 1923	Lieutenant Commander Mason, flying a DH-4B, makes the first successful one-day flight between Pensacola and Washington, DC.
September 20, 1923	First use of U.S. Navy seaplanes to spot gunfire for Fort Barrancas' Battery Pensacola target practice.
January 15, 1924	First non-stop commercial flight between Pensacola and New Orleans departs on Gulf Coast Air Line.
March 1924	Charles Lindbergh, while still an unknown, lands his JN-4C at NAS Pensacola while traveling from St. Louis to Brook Field to enter U.S. Army Air Service flight training.
February 7, 1927	New landing field west of Pensacola offered to the U.S. Navy to replace Corry Field, which is now overrun by city development.
February 17, 1927	New landing field accepted by the U.S. government.

April 24, 1927	First visit of a rigid airship, the USS LOS ANGELES (ZR-3), to Pensacola.
April 1927	NAS Pensacola seaplanes fly in support of Mississippi flood relief.
May 20/21 1927	Charles Lindbergh flies the Atlantic.
July 9, 1927	New landing field west of Pensacola is officially dedicated and retains name of Corry Field. Original landing field now known as Old Corry Field.
July 27, 1927	Airmail plane lands at Old Corry Field on experimental trip to determine the practicality of a permanent airmail line on the Gulf Coast.
October 9, 1927	Charles Lindbergh visits NAS Pensacola during his tour of the nation.
January 6, 1928	Mabel Cody's Flying Circus arrives at Old Corry Field.
April 19, 1928	Pensacola begins use of airmail using a railway connection from Pensacola to Mobile to connect to aeroplane service.
November 19, 1928	A barnstorming Ford tri-motor of Tri-Motored Air Tours lands at Old Corry Field.
February 24, 1929	Pensacola Lions Club announces their main objective for 1929 is to establish a Pensacola municipal airport.
March 15, 1929	First airmail originating from Pensacola occurs when severe flooding strikes Pensacola, cutting off all access. U.S. Navy seaplanes fly mail to Mobile and New Orleans and this mail is stamped "Air Mail via ___ " with a U.S.N. postmark date.
April 1929	Pensacola's first flying club, chartered by the American Society for the Promotion of Aviation, is established.
September, 1929	Work begins on Pensacola Municipal Airport established on a leased 100-acre section of Old Corry Field.
October 1929	Auxiliary landing field "Zee" established near the Lillian Bridge road.
October 1, 1929	NAS Pensacola Pigeon Station decommissioned.
October 27, 1929	Charles J. Weger announces his intentions to organize a company to be known as the Pensacola Metal Aircraft Corporation to build huge civilian seaplanes in Pensacola.

CHAPTER 1
HISTORY OF PENSACOLA

Figure 1. Drawn in the late 1760s, this is the oldest known image of Pensacola at its present location. (Courtesy West Florida Historic Preservation, Inc.)

PENSACOLA ESTABLISHED

While several early 15th century Spanish explorers are known to have visited the body of water that is now known as Pensacola Bay, it was Don Tristan de Luna who arrived here on August 14, 1559 to attempt the first European colony in the United States. He named Pensacola Bay Bahia de Santa Maria Filipina de Ochuse and settled at a now unknown location. Devastated by a hurricane that sank most of his ships and decimated by the harsh living conditions, this colony lasted for only two years before it was abandoned. More than a hundred years passed before any Europeans returned.

On November 21, 1698 Don Andres de Arriola arrived to establish the first permanent settlement in Pensacola. Located on the present grounds of Naval Air Station Pensacola, it was known as Presidio Santa Maria de Galve and included a wooden fort named Fort San Carlos de Austria.

The French captured this site in 1719 but Pensacola was returned to Spanish control by treaty in 1722. The Spanish moved the settlement to Santa Rosa Island and built Presidio Isla de Santa Rosa, located just east of the present location of Fort Pickens on Santa Rosa Island.

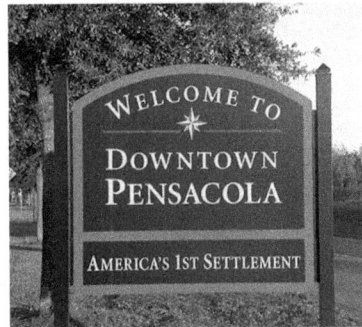

Figure 2. St. Augustine, Florida is recognized as the oldest *continuously* occupied European settlement in the United States by virtue of the permanent Spanish settlement established there in 1565. This was six years after Pensacola's own ill-fated Spanish settlement attempt in 1559. (Courtesy of author)

PENSACOLA'S FORTS

Beginning in 1698 a series of Spanish and British fortifications were constructed in and around Pensacola. By the time the United States acquired Pensacola in 1821 nearly all of these facilities were in complete disrepair and a major construction effort was initiated to defend the entrance to Pensacola Bay, safeguard the city of Pensacola, and protect the Pensacola Navy Yard. Between 1829 and 1859 three forts and a redoubt were built while an existing water battery was restored.

Fort Pickens, the largest of the forts, is located on Santa Rosa Island on the eastern side of the channel into Pensacola Bay and was built between 1829 and 1835.

Fort McRee, which was located on Foster's Bank on the western side of the channel into Pensacola Bay, was built between 1834 and 1837. Severely damaged during the Civil War, the fort's final casements collapsed into the Gulf of Mexico during a storm in July 1931.

Fort Barrancas, which is located on a bluff directly north of the entrance into Pensacola Bay, was built between 1839 and 1844 and included a post to house the fort's garrisons, storehouses and workshops.

Advanced Redoubt, which is located inland approximately three-quarters of a mile north of Fort Barrancas, was built between 1845 and 1859 to protect Fort Barrancas and the Pensacola Navy Yard against land attack.

Battery San Antonio, also known as the Water Battery or Fort San Carlos, is located directly in front of Fort Barrancas and was originally built in 1797 and restored from 1839 to 1844. A water battery is designed to ricochet shots off the water.

During the timeframe of this book only Fort Pickens and Fort Barrancas remained in service under the command of the U.S. Army's 13th Coast Artillery Regiment headquartered at Fort Barrancas. Several artillery batteries were operational, including Battery Pensacola with two 12-inch cannons mounted on disappearing carriages, located in the center of the Fort Pickens. After World War II all of these fortifications were declared surplus and transferred to the National Park Service where they are maintained today as tourist attractions.

Figure 3. The value of this undated photograph is its superb illustration of the size of the Fort Barrancas complex. Fort San Carlos is at the bottom right with Fort Barrancas to its left. Behind both is Fort Barrancas' Post with its support facilities and large parade deck. (Courtesy Pensacola Historical Society)

After a 1752 hurricane ruined this settlement, the Spanish eventually relocated to the fourth and present location of the city of Pensacola and named their town Presidio San Miguel de Panzacola.

Control of this desirable deep water port city on the Gulf of Mexico shifted back and forth several times between Spain, France and Great Britain with the city's name gradually evolving into Pensacola. On July 17, 1821 permanent possession of Pensacola transferred to the United States as Spain ceded Florida according to the terms of the Adams-Onis Treaty. During the Civil War, the Confederate States of America occupied Pensacola for a brief period, making their flag the fifth to fly over this city, a source of great pride to Pensacolians celebrated in their self-description as the City of Five Flags.

PENSACOLA NAVY YARD ESTABLISHED

In September 1825 the Secretary of the Navy was looking for a more suitable homeport for the U.S. Navy's West India Squadron, a flotilla of warships combating pirates in the Caribbean and the Gulf of Mexico. Accordingly, he directed three senior Naval officers to proceed to the Gulf of Mexico and select a site for the establishment of a navy yard and depot.

On October 25, 1825, the United States Sloop of War HORNET "...eleven days from the Capes of Virginia..." passed over the sand bar at the mouth of the Bay of Pensacola, slipped by the ruined forts that once guarded her entrance and anchored abreast Pensacola to a warm welcoming committee.[1] Embarked were Commodore Lewis Warrington, Commander of the West India Squadron, Commodore William Bainbridge, and Captain James Biddle. After several days of surveying, they selected an area surrounding Tartar Point, which was located on the western Pensacola Bay seashore to the east of Fort Barrancas and seven miles southwest of the city of Pensacola.[2] Their

Figure 4. 1881 chart showing Fort Pickens, Fort McRee, Fort Barrancas, and the Pensacola Navy Yard. Not shown is the Advanced Redoubt, located north of Fort Barrancas, and Pensacola, located northeast of the Pensacola Navy Yard. (Author's collection)

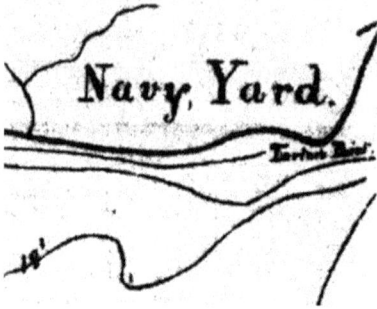

Figure 5. Tartar Point, selected in 1825 for the site of the Pensacola Navy Yard and present home to Naval Air Station Pensacola, was originally used as a sighting point by mariners transiting the Pensacola Bay channel. (Courtesy University of West Florida Special Collections Library)

report was submitted on November 4, 1825, favorably endorsed by the Secretary of the Navy on December 2, 1825, and approved by President John Quincy Adams on December 2, 1825.

Of interest, a history of the U.S. Marine Corps in Pensacola states that on November 6, 1825 "Seven Marines from the frigate John Adams were transferred ashore as a guard for the Navy Yard."[3]

Construction of the Navy Yard, including a stone wharf, brick storehouses, quarters, and workshops slowly moved forward. However, by 1850 an observer noted "It is now the largest and most beautiful Yard in the country. Handsome dwellings, substantial storehouses and really elegant workshops arise, month and month, with great rapidity."[4] Nonetheless, further work continued intermittently over the next several decades as the U.S. Navy sought to establish a "…first-class navy yard and depot for the construction, repairing, and provisioning of ships of the fleet."[5]

The Pensacola Navy Yard eventually encompassed nearly 80 acres surrounded by two tall brick walls to provide protection from a land attack. The first wall extended north from Pensacola Bay to where it joined a second wall extending east to the bay. No walls were built along the waterfront. Most of the walls remain intact and visible to this day.

Figure 6. View of the Pensacola Naval Yard in 1885. Most of these buildings are still in existence. (Courtesy Pensacola Historical Society)

Figure 7. 1916 view looking beyond the north wall of the Pensacola Navy Yard into the village of Woolsey. (Courtesy T.T. Wentworth, Jr. Collection, West Florida Historic Preservation, Inc.)

WARRINGTON AND WOOLSEY ESTABLISHED

Due to the distance from the Pensacola Navy Yard to the city of Pensacola, a significant housing problem presented itself for the construction workers and civilian yard employees. As a solution, the U.S. Navy leased lots outside the brick walls of the shipyard to the north and west of the Navy Yard for the workers to build homes and businesses at their own expense. Still on Naval reservation property, it was clearly understood by all concerned that these leases could be "...summarily terminated by the commandant.... without recourse."[6]

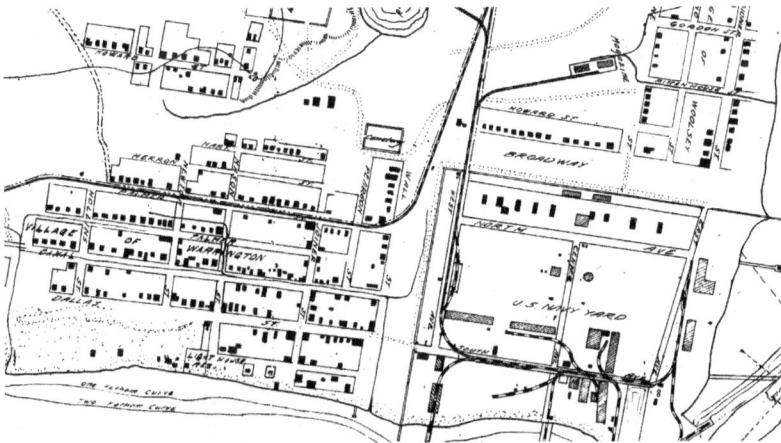

Figure 8. A very detailed map dated 1916 showing the buildings of both the village of Warrington, located to the west of the old Pensacola Navy Yard's brick walls, and the village of Woolsey, located to the north. (Courtesy U.S. Navy)

Figure 9. 1917 view looking beyond the west wall of the Pensacola Navy Yard into the village of Warrington. (Courtesy Pensacola Historical Society)

Therefore, two villages were created. Warrington, named for Captain Warrington, the first commandant of the Pensacola Navy Yard, was located on the western side of the Navy Yard wall; and Woolsey, named for Captain Woolsey, the second commandant of the Pensacola Navy Yard, was built on the northern side. Homes, businesses and recreation areas flourished in both of these villages. Many early pictures of the Pensacola Navy Yard, Warrington, and Woolsey make them appear to be barren, sandy places. However,

Figure 10. The Pensacola Electric Company's Bay Shore Line, shown here departing the Navy Yard in 1909, provided transportation between Pensacola, Woolsey, the Pensacola Navy Yard, Warrington and Fort Barrancas. (Courtesy Pensacola Historical Society)

former residents remembered them as beautiful spots, well kept, with many of the homes having fruit trees, flower and vegetable gardens, all bordering on white beaches.[7]

In March 1919 a report noted that "The villages of Warrington and Woolsey contain together about 1500 people, Navy employees and their families, who are allowed to own houses on the Naval Reservation."[8] As will be discussed in detail later, within a few years the U.S. Navy would require the land both villages occupied. Woolsey would be completely razed and Warrington moved to its present location to accommodate the expansion of Naval aviation activities beyond the walls of the Pensacola Navy Yard.

TRANSPORTATION

In 1921, the Commandant of Naval Air Station Pensacola wrote "…our station is somewhat off the beaten path in regard to the rest of the navy, and 'isolated' in a Naval sense."[9] The same could be said for its local geographic position. Since only unpaved, sandy trails linked the city of Pensacola with the station, the eight-mile distance between the two presented a formidable journey. Access to reasonably priced automobiles, and paved roads would not occur for decades, so railways provided the first mass transportation between Pensacola and Woolsey, Warrington, the Pensacola Navy Yard and Fort Barrancas. Initially, the Pensacola Terminal Company operated a "Dummy Line"* steam engine streetcar railway system between the sites.[10] Operated along the bayshore, in 1903 this line was succeeded by the Pensacola Electric Company and converted to electric streetcars. Known as the Bayshore Car Line, service was discontinued in 1932 and replaced by the buses of the Pensacola Coach Corporation.

Inside the Pensacola Navy Yard, U.S. Navy locomotives hauled freight between the various buildings and through the northern wall to the brick enclosed Powder Magazine in Woolsey. Railroad service was established prior to the Civil War and cars from Pensacola carrying material for the station were brought to a

Figure 11. A small bit of evidence of a railroad system within the gates of the Pensacola Navy Yard and Naval Air Station Pensacola are these partially uncovered tracks visible along West Road. (Courtesy of author)

siding in Warrington where they were transferred to the station's locomotive.[11] Eventually over "…seven and a quarter miles of track wound its way around the base."[12]

PENSACOLA NAVY YARD IN DECLINE

Over the years the fortunes of the Pensacola Navy Yard ebbed and flowed with strategic requirements, politics, wars and expeditions, severe hurricane damage, shipbuilding and repair contracts, yellow fever epidemics, and the availability of maintenance, repair and construction funding. By the late 1800s Naval activity at the yard had diminished to the point where it was placed in an inactive status with occasional flurries of ship visits, repairs and construction activities in the early 1900s. A September 1906 hurricane devastated the Yard and the damage was never fully repaired.

Matters were made worse in 1909 when Secretary of the Navy George von L. Meyer, serving under Republican President William Howard Taft, sought to close the southern navy yards in a cost saving economy and reform program. After inspecting the Pensacola Navy Yard on October 26, 1910,[13] Meyer recommended to Congress that Pensacola, New Orleans and several other yards be closed.[14] The Pensacola News Journal accused New Englander Meyer of practicing "partisan politics and sectional politics" for his protection

---

* This was called the Dummy Line as the locomotive's steam boiler was enclosed within a streetcar facade. According to theory, this would avoid frightening horses, as all the cars would appear the same. It must not have traveled very fast as one local recalled a saying "If you leave downtown at half past one, you'll get to the Navy Yard at the setting sun."

of the northern naval yards at the expense of the southern shipyards.[15] It was as if economic, environmental and political forces were conspiring to rob Pensacola of her Naval forces.

The situation at Fort Barrancas, however was not quite so bleak. In fact, two U.S. Army officers had enough time and resources to experiment with an entirely new technology.

NOTES

[1] Arrival of the Hornet. (1825, October 29). *Pensacola Gazette and West Florida Advertiser.*

[2] Pensacola: Copy of a letter from the Hon. J.M. White (1826, January 7). *Pensacola Gazette and West Florida Advertiser.*

[3] Pensacola Marine Barracks One of Oldest in the Corps. (1955, November 10). *Gosport.*

[4] Bits Of History. (1929, May 17). *Air Station News, Pensacola, Florida.*

[5] Pearce, G.F. (1980). *The U.S. Navy in Pensacola from Sailing Ships to Naval Aviation (1825 – 1930).* Pensacola, FL: University Presses of Florida, page 16.

[6] Pearce, G.F. (1980). *The U.S. Navy in Pensacola from Sailing Ships to Naval Aviation (1825 – 1930).* Pensacola, FL: University Presses of Florida, page 16.

[7] Nix, J.R. (n.d.). *Shadows of the Past of Pensacola Navy Yard, Old Warrington, Woolsey, Fort Barrancas, Fort Pickens, Fort Redoubt As I Knew Them.* No publisher.

[8] Bennett, F.M. (1919, March 15). *Historical Information, Collection of.* Special Collections, 73-6, Box 2, John C. Pace Library.

[9] From the Commandant. (1921, August 5). *Air Station News,* Pensacola, Florida.

[10] Pitts, J. (1977, June 5). The Pensacola Street Railway System. *Pensacola News Journal.*

[11] Station Railroad Keeps Freight Flowing. (1958, August 15). *Gosport.*

[12] This is Perhaps Oldest Railroad in Naval Service. (1958, August 15). *Gosport.*

[13] Secretary of Navy Meyer Here on Inspection Tour. (1910, October 27). *The Pensacola Journal.*

[14] Abolition of Navy Yards is Recommended. (1910, December 6). *The Pensacola Journal.*

[15] Sec'y Meyer's Report and The Pensacola Navy Yard. (1910, December 6). *The Pensacola Journal.*

CHAPTER 2
THE RHODES AEROPLANE

Figure 1. On February 6, 1910 this photograph appeared on the front page of The *Pensacola Journal* showing Lieutenant Rhodes'and Major Gossman's aeroplane at Fort Barrancas – the first known aeroplane to fly "near" Pensacola. (Courtesy *Pensacola News Journal*)

U.S. ARMY AEROPLANE AT FORT BARRANCAS

On November 14, 1909 slightly less than six years after the Wright brothers flew at Kitty Hawk, and more than four years before the arrival of Naval aviation at Pensacola, an astonishing story appeared on The *Pensacola Journal's* front page. Unknown to almost everyone in Pensacola, two enterprising officers, Lieutenant Albert Rhodes of the 13th Coast Artillery Corps and Major George Gossman, a doctor with the U.S. Army Medical Corps, were exploring the concept of aeromedical evacuation. The article is quoted in its entirety:

Figure 2. Unfortunately this photograph was not annotated, but standing in front of Pensacola's first aeroplane appears to be Lieutenant Rhodes on the right, Major Gossman in the middle and an unidentified officer on the left. (Courtesy Pensacola Historical Society)

LIEUT. RHODES CONSTRUCTING AN AEROPLANE AT BARRANCAS

Army Officer Will Have His First Model Ready For Test in Six Weeks

HAS BEEN QUIETLY AT WORK FOR THE PAST SIX MONTHS BUILDING A MACHINE ALONG DIFFERENT LINES FROM ANY OF THOSE OF OTHER FLYING MACHINES – CAPTAIN GOSSMAN, WHO IS ASSOCIATED WITH THE BUILDER IS IN WASHINGTON

Figure 3. Major George Gossman on the left and Lieutenant Albert Rhodes on the right, sitting in their aeroplane. (Courtesy *Pensacola News Journal*.)

Quietly at work for many months and not even known to many of the officers at the post, Lieut. A. L. Rhodes of Fort Barrancas, has under construction, and in fact, nearing completion an aeroplane which, when completed, will be along lines different to those built by the Wrights, Curtiss, and other aerial navigators. The machine on which Lieut. Rhodes has been quietly though diligently working and studying, will be completed within the next six weeks and then the inventor will make his first test.

"Yes, I am building an aeroplane," guardedly admitted the army officer to a Journal representative yesterday afternoon. "I have been at work on it for some time now and hope it will be a success when tested, but of course that remains to be seen. Dr. Gossman, who is associated with me in the building of the machine, is now in Washington, where he has gone for examination for promotion, and while there he hopes to have patented some of my ideas on aeroplane construction.*

DISTINCTIVE FEATURES

The machine was designed by Lieut. Rhodes and in it he has incorporated quite a number of distinctive features, and as far as known, not contained in any other machine now in use or designed. One of the features will be an endeavor to make the machine soar and to bring this about there are four ailerons at the end of each plane or wing. This, it is believed, will give it lateral stability. The idea originated with Lieut. Rhodes after he had studied for weeks the manner in which the buzzard soars high in the air and apparently without any effort whatever. In fact the lieutenant has followed, to a great extent in building his machine, the buzzard.

ITS CONSTRUCTION

The frame of the machine, with which Lieut. Rhodes hopes to solve the problems of aerial navigation, is being constructed of oak and juniper and braced with copper wire. The wings will be 26 feet from tip to tip, it will have 250 square feet of supporting surface and will weigh in the neighborhood of 350 pounds. Every piece of material is carefully tested before it is placed, and the builder is especially careful in his selection of the juniper. The horsepower of the motor has not yet been decided upon, as the first test will be made by towing with an automobile. By this means the motor necessary can be ascertained with a degree of satisfaction which could not otherwise be done. When the test is made the machine will be mounted upon three bicycle wheels and the trial will be on the new cemetery road, a long tow line being fastened to the automobile and attached to the machine.

A CLOSE STUDENT

That Lieut. Rohdes (sic) is a close student of aerial navigation is clearly shown by his conversation. The early attempts at flying machine construction, the manner of construction of the Wrights, Curtiss and other aeroplanes, what features cause their success or failure and why a machine constructed along the lines that he has mapped out should prove a success.

He and Dr. Gossman hope to interest the army in their new machine, at least to the extent of getting some encouragement, and while in Washington Dr. Gossman will go before the department and see if this cannot be secured.

Lieut. Rhodes is the only officer in either the army or navy who has attempted the construction of an aeroplane and those officers who are aware of the inventor's plans believe that it will be a great success.[1]

TEST FLIGHT

On January 25, 1910 they made their first test flight without a motor. As described in *The Pensacola Journal:*

The test was made behind a big touring car, which supplied the power. A long hawser was attached to the automobile. Lieut. Rhodes was seated behind the car for a short distance and then, as the auto speeded up, the aeroplane lifted itself from the ground and soared along behind the automobile, remaining perfectly steady and not giving its inventor the least trouble in guiding it.[2]

Lieutenant Rhodes ordered a motor for his aeroplane and received it in February 1910.[3] Apparently the motor had problems generating enough power and it was worked upon and tested for several months. On June 18, 1910 Lieutenant Rhodes attempted three takeoffs in which he ran the engine at full speed while being towed behind Major Gossman's automobile. Each time the aeroplane failed to lift off the ground. After further tests they attempted their first motorized flight and the aeroplane rose to a "...100 to 150 feet height for 500 yards until an oil line broke and the plane wrecked in a tree."[4] Lacking personal funds to continue their experiments, Major Gossman approached the U.S. Army for money but he was rejected and no further aviation activity was conducted at the fort.

Since the aeroplane never left the confines of Fort Barrancas, the citizens of Pensacola would have to wait several more years to see the first flight of an aeroplane over their city.

Figure 4. The Fowler Flying Machine constructed by John Fowler in Mobile before 1900. However, plagued by the lack of an adequate engine, like all early aviators, there is no documented evidence that he ever flew. (Courtesy Mobile Public Library)

Great Southern Aviation Meet
at Mobile

| Six Days | **NOVEMBER** 21, 22, 23, 24, 25, 26 | Six Days |

COME

Reduced Rates on All Railroads

Two of Glenn H. Curtiss' most daring Birdmen will make daily flights and aerial exhibits at the Head of Government Street every afternoon week of November 21st in Curtis aeroplanes under auspices of

THE MOBILE REGISTER

MOBILE, ALABAMA

Pensacola's rival, nearby Mobile, was a surprising hotbed of early aviation activity. One Mobilian, an inventor by the name of John Fowler, had constructed aeroplanes as early as 1893, although they never flew.

Someone who did fly over Mobile was Mr. J.A.D. McCurdy. On November 22, 1910 working for the Curtiss Exhibition Company, he made Mobile's first heavier-than-air flight. Joined by Augustus J. Post, they brought three flying machines to the "Great Southern Aviation Meet" hosted by Mobile and thrilled thousands of spectators with dozens of flights over the city. One of the flying machines was the "Hudson Flier," the aeroplane that Glenn Curtiss flew from Albany to New York to win the final leg of the Scientific American Trophy.

Besides a spectacular crash by McCurdy, who emerged unhurt, the absolute highlight of the meet was the appearance of Eugene Ely, who only eleven days earlier had made the first take-off from a warship at sea aboard the USS BIRMINGHAM. During one exhibition flight, both McCurdy and Ely were airborne simultaneously, providing Mobilians a spectacular sight when they flew "...one above the other in the same direction."

The grand finale of the aviation meet occurred when McCurdy blew up a battleship on the final day, simulated by the dropping of oranges on canvas squares laid out on the field.

NOTES

[1] Lieut. Rhodes Constructing an Aeroplane at Barrancas. (1909, November 14). *The Pensacola Journal*.

[2] First Flight of the Rhodes Airship Highly Successful. (1910, January 26). *The Pensacola Journal*.

[3] Motor For New Airship Comes. (1910, February 2). *The Pensacola Journal*.

[4] Futrell, R.F. (1961). *Development of Aeromedical Evacuation in the United States Air Force 1909-1960*. Maxwell AFB, AL: Air Force Historical Research Agency (USAF Historical Studies: No. 23).

CHAPTER 3
THE NASSR AEROPLANE

Figure 1. No photograph of the second aeroplane to appear in Pensacola is known to exist. Built by Tony Nassr in 1911, his attempted flight from Kupfrian's Park on June 13, 1911 ended in disaster. (Courtesy Pensacola Historical Society)

THE NASSR AEROPLANE AT KUPFRIAN'S PARK

Mr. Anthony N. (Tony) Nassr, an accomplished aeronaut known as the "Daring Syrian," first arrived in Pensacola on October 25, 1910 to provide lighter-than-air exhibition flights for the Pensacola Interstate Fair.[1] He made several successful ascensions and upon completion of the fair he announced that he would spend the winter here "...constructing and experimenting with heavier-than-air machines."[2]

Leasing the "Intendencia street airdrome" for his workshop, he first built a working model of his design before beginning construction on the actual flying machine. His aeroplane was described as:

...thirty feet in width with a five-foot double surface, the planes being five feet apart. The machine will be of the biplane type and will have, when completed, combined features of the Curtiss and Wright machines. However, it will be distinct in some features. Its surface will be double curvature. The total weight will be from four hundred and twenty-five to five hundred pounds, exclusive of the operator. The engine which Mr. Nassr has chosen will have a capacity of from twenty-five to thirty horse power of the two cylinder, double opposed, air cooled and break test type.[3]

Mr. Nassr placed an order for an engine with the Detroit Aeroplane Company.[a] The Aeromotor, as it was known, had "...attracted considerable attention with its clean-cut, graceful lines and finished appearance that immediately appeals to the observer and with points of construction that call forth the admiration of engineers."[4] Upon its arrival in February 1911, the engine was displayed, with propeller attached, for several days in the window of *The Pensacola Journal* to the admiration of several hundred spectators. Visitors also were welcomed to visit his workshop to view the actual aeroplane. Construction continued on the aeroplane for several more weeks until May 5, 1911, when to the amazement of Pensacolians who had never seen an aeroplane, the craft was moved to Kupfrian's Park[b] for final preparations.[5]

On June 13, 1911 Mr. Nassr attempted his first flight in his new aeroplane. As described in *The Pensacola Journal:* "Going at a speed of thirty-five miles an hour Aviator A.M. Nassr, in an aeroplane which he had just completed, collided with a stump and the result is that the aviator received slight bruises and the machine sustained damages amounting to several hundred dollars."[6]

The article went on to note that:

The machine worked beautifully and during the preliminary tests bobbed up and down like a huge bird and when the accident happened the machine was several feet off the ground. Only a few persons were present to see the preliminary flights and when they were remarking upon the success of the trial the collision came, it making a terrific noise. They rushed to where the machine had fallen expecting to find the aviator mutilated but he was smiling and looking at the machine that he had just felt rise from the ground with him as easily as if a real bird.[7]

Mr. Nassr immediately ordered replacement parts and felt confident that he could make a successful flight once he found a more suitable location free of tree stumps. After rejecting several more sites due to stumps or soft ground, nothing further was reported on Mr. Nassr and it is assumed that he departed the city to conduct his experiments elsewhere.[8]

Mr. Nassr went on to command an air defense base in New York during World War I, serve as an U.S. Army balloon inspector at the Goodyear plant, and later become the director of the first airport in Toledo, Ohio.[9]

However, Pensacolians were once again left waiting to see the first flight of a heavier-than-air machine over their city. Surprisingly, it was not long in coming.

---

[a] While waiting for his engine, Mr. Nassr provided dirigible exhibition flights for several nearby towns.

[b] Kupfrian's Park was located in the northwest section of Pensacola near the present Baptist Hospital.

FLORIDA'S AVIATION HISTORY

Florida's early aviation history begins with lighter-than-air flights in Jacksonville where a manned balloon of unknown origin was reported to have drifted over the city in 1878. Lincoln Beachey, shown in his Curtiss Pusher, made the first airship flight in Florida's history at Jacksonville in December 1908.

In February 1910, Beachey made the first heavier-than-air flight in Florida's history at the Orange County Fair in Orlando. Competing with three other aviators for a $1,500 prize to the aviator who could stay aloft for five minutes, he was the only contestant to get airborne and easily surpassed the required time. In March 1911 at Tampa, Beachey also made the first night flight in Florida's history, completing his hat trick of Florida aviation firsts.

In 1910 Edward Andrews of Chicago built the world's first twin-engine aeroplane at Daytona Beach. Unfortunately, it crashed shortly after take-off.

The first west-to-east transcontinental flight in United States history terminated at Jacksonville Beach on February 12, 1912. Flying his Wright aeroplane *Vin Fiz*, Robert Fowler took 115 days to complete his journey.

The first woman to fly in Florida was Ruth Law who flew her Wright B on January 12, 1913 at Daytona Beach. Earning as much as $9,000 ($126,000 in 2002 dollars) a week in exhibition flying, she also was the first woman to loop an aeroplane and fly at night.

Florida is perhaps most well known as home to the first American and international airlines. On January 1, 1914 the St. Petersburg-Tampa Airboat Line inaugurated the world's first scheduled commercial airline, flying a two-person flying boat between Tampa and St. Petersburg. On November 1, 1920, Aeromarine Airways began the world's first scheduled international airline service with eight person flying boats between Key West and Havana, Cuba.

NOTES

[1] Air Ship Came to Pensacola Last Evening. (1910, October 25). *The Pensacola Journal.*

[2] Pensacola To Be the Home of Aeroplane. (1910, December 16). *The Pensacola Journal.*

[3] Construction of Aeroplane to be Rushed. (1911, January 10). *The Pensacola Journal.*

[4] Nassr Buys Latest Make Aero Engine. (1911, February 22). *The Pensacola Journal.*

[5] Nassr Moves Machine to Kupfrian Park. (1911, May 5). *The Pensacola Journal.*

[6] Smashes His Aeroplane On A Stump. (1911, June 14). *The Pensacola Journal.*

[7] Smashes His Aeroplane On A Stump. (1911, June 14). *The Pensacola Journal.*

[8] Seeking For Location For Trial Flight. (1911, July 8). *The Pensacola Journal.*

[9] Tony Nassr. Retrieved from http://www.toldeblade.com/apps/pbcs.dll/article?date=200003 26&category=sratole.

CHAPTER 4
THE NELSON AEROPLANE

PENSACOLA AVIATION MEET
Exhibition Flights at Palmetto Beach
Thanksgiving Afternoon, Nov. 30—Sunday Afternoon, Dec. 3

N. J. NELSON one of the world's best aviators in thrilling aeroplane flights

First Airship Flight in Western Florida. 2 Flights Each Afternoon Nov. 30, Dec. 3

Figure 1. Although an aviation meet of only one aeroplane and one aviator, the honor of the first person to make a heavier-than-air flight over Pensacola is given to Mr. Nels J. Nelson. (Courtesy *Pensacola News Journal*)

PENSACOLA AVIATION MEET

With Pensacola known worldwide as the "Cradle of Naval Aviation" or "Annapolis of the Air," one would logically conclude that a Naval aviator finally achieved the distinction of making the first successful aeroplane flight over this fair city. However, the honor of making the first heavier-than-air flight over Pensacola befell not to a Naval aviator, Army airman, or barnstorming aeronaut. Instead, an exhibition pilot, who was performing in nearby Mobile, was actually paid to come to Pensacola to achieve this small modicum of local fame.

Exhibition flying was created by the two great competitors of early commercial aviation, the Wright brothers with their unique 'Flyers' and Glenn Curtiss with his equally innovative 'Pushers.' Each formed their own company and their own cadre of exhibition flyers to promote their product. Some of the first purchasers of their aeroplanes were entrepreneurs who traveled across the country entertaining local communities with lucrative aviation exhibitions.

Pensacola was bypassed on this entertainment circuit until November 18, 1911 when the *Pensacola Evening News* received an unexpected telegram. An agent of the Mills Aviators, who were providing exhibitions flights at a fair in Mobile, wired:

Will come to Pensacola Monday to arrange for airship exhibitions Thanksgiving after-noon if you think conditions favorable. Answer Cawthon Hotel. Have you had any flying machines?[1]

Contracts, arranged through "several prominent young men of the city," were signed for an appearance during the Thanksgiving holiday.[2] A letter from the Gulf Coast Tropical Fair Association, the group in Mobile who was hosting the Mills Aviators, certified that Pensacolians would not be disappointed. They described Nelson's flights as "...beautifully made, averaging 600 feet and covering a space of six miles in each flight, in circles. In one of their flights he inscribed a figure eight, which was very attractive. The flights of Mr. Nelson increased our attendance more than 500 per cent."[3]

The event was heralded as the "Pensacola Aviation Meet" and it consisted of exactly one pilot, Mr. Nels J. Nelson, flying precisely one aeroplane, a Curtiss-Mills Pusher. The meet was held at Palmetto Beach Park, located a few miles west of the city on the bayshore. Extra streetcars to the park were added to accommodate the thousands of spectators expected to observe the first aeroplane flight in Western Florida's history. Billboards promoted "N.J. Nelson, of the Mills World's Famous Birdmen, a Dare-Devil." Admission price was set at 50 cents for adults ($10 in 2002 dollars).

Mr. Nelson, accompanied by his assistant Richard Nygren and James R. Mills of the Mills Aviators, arrived at the San Carlos Hotel on the evening of November 28, 1911.[4] The next day they assembled their aeroplane, which was shipped from site to site rather than flown, tested their motor, built a tent hangar and pronounced the field favorable for flight operations.[a] Mr. Nelson was described as being very courteous to spectators, patiently explaining every part of his aeroplane in detail.

FIRST AEROPLANE FLIGHT OVER PENSACOLA

On November 30, 1911 Pensacolians awoke to the coldest Thanksgiving Day in their city's history, the first of two records set this date. For in the afternoon, with ideal flying conditions of crisp cool air and light westerly winds, a "Man-Bird" made the first heavier-than-air flight in Pensacola's history. As described in *The Pensacola Journal*, Mr. Nelson and his aeroplane *Eagle* presented Pensacola with an aerial dessert for their Thanksgiving Day dinner:

---

[a] Their aeroplane was identified as a Curtiss-Mills version of the Curtiss Pusher, reflecting some non-specific improvements made by the Mills brothers to enable "quick get-aways and quick climbing."

Figure 2. This non-annotated photograph found in the Pensacola Historical Society's collection may be of Nelson's aeroplane on display at Palmetto Beach. (Courtesy Pensacola Historical Society)

About 3 o'clock he had his machine wheeled out on Palmetto Beach; it took about six men to push it through the thick soggy sand, and they seemed very much relieved when Nelson signaled them that it was in the right position. He then mounted the aviator's perch, and with one turn of the propeller by his assistant, Richard Nygren, he was off. He skidded off over the wet sands and gradually lifted up and up, over the heads of the awe-stricken spectators, some of which glimpsed for the first time in all their lives this much-talked-of-modern-day miracle. He reached an altitude of about five hundred feet on his first turn and his manouevres (sic) for a while made many a heart thump. They seemed very much relieved when he at last descended, amid the applause of all.[5]

Nelson flew for two more days before departing on a world tour. However, nearly three years passed before Pensacolians saw their second successful aeroplane flight.

NOTES

[1] Pensacola To See Men Fly Thanksgiving. (1911, November 18). *Pensacola Evening News*.

[2] Biplane Will Be Seen Here. (1911, November 21). *The Pensacola Journal*.

[3] An Aviator of Wide Experience is Nels Nelson. (1911, November 27). *Pensacola Evening News*.

[4] Nels Nelson, the Aviator, Reaches City. (1911, November 29). *The Pensacola Journal*.

[5] Fine Flights Witnessed By Many People. (1911, November 30). *The Pensacola Journal*.

CHAPTER 5
NAVAL AVIATION COMES TO PENSACOLA

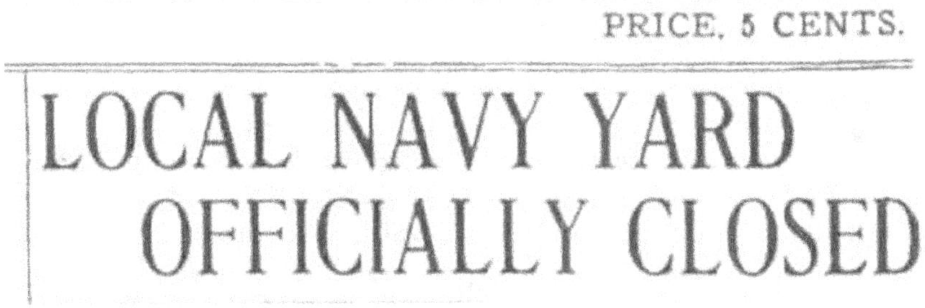

PRICE, 5 CENTS.

LOCAL NAVY YARD
OFFICIALLY CLOSED

Figure 1. October 21, 1911 headline from *The Pensacola Journal* announcing the closure of the Pensacola Navy Yard. (Courtesy *Pensacola News Journal*)

PENSACOLA NAVY YARD CLOSED

None of these remarkable aeronautical events deterred the seemingly inevitable fate of the Pensacola Navy Yard. On July 1, 1911 orders were received to lay off nearly all of the 190 remaining workers at the yard.[1] A few days later all stores and equipment were ordered packed for immediate shipment to other yards.[2] A few months later the Naval collier STERLING was dispatched to Pensacola to collect these goods and on October 19, 1911, she departed with more than a million pounds of material.[3]

On October 20, 1911 the Pensacola Navy Yard was officially closed, but not completely abandoned as a "...chief carpenter will remain as the officer in charge of the yard with three watchmen to look after the machinery and other property."[4] A wireless station also was kept in operation at the yard.[5]

As it turned out, nearly everything at the station was expendable. In November 1911 two "mammoth sea-going naval tugs" arrived to tow a half dozen or more coal barges used at the Pensacola Navy Yard to Guantanamo Bay, Cuba.[6] *The Pensacola Journal* accused Secretary Meyer of dismantling the Navy Yard in stages and to "...remove every article of value that is removable to his more favored New England or Cuban stations..."[7]

In April 1912 the collier STERLING returned to strip the Pensacola Navy Yard of everything useable, including building materials, furniture, and boats.[8] What couldn't be loaded was auctioned off, including "...a locomotive, several flat cars, scrap iron and copper, a horse, vehicles and material which cannot be profitably moved."[9]

Even the lights were turned off. Captain W.G. Barrow, master of the steamer *Tarpon*, told *The Pensacola Journal* that "...he had been sailing the waters of Pensacola bay since 1868 and this is the first time within his memory when he has not seen lights along the landings at that place."[10]

Although there would be several ship visits over the next few years, the Pensacola Navy Yard was at the absolute nadir of its existence.

PENSACOLA COMMERCIAL ASSOCIATION

Although not unexpected given the decline in activity at the yard and Secretary of the Navy Meyer's warnings, Pensacola still appeared to be stunned into inaction by the closure. It was not until June 1912 that the Pensacola Commercial Association stepped forward to formulate a plan of action to secure Congressional support to reopen the yard.[11] The Commercial Association, an organization of Pensacola businessmen formed to advance the city's professional interests and presided over by Mr. C.E. Dobson, proved to be indefatigable in their lobbying efforts to convince the U.S. Navy to reopen the Pensacola Navy Yard.

PENSACOLA HEADS TOWARD RECESSION

The closing of the Navy Yard was the harbinger of future misfortunes for Pensacola. While 1912 and 1913 were exceptional years in terms of exports and imports through the city's port, Pensacola's economic export base of fishing, lumber, cotton and naval stores (rosin and turpentine) was soon in steep decline.[12] Pensacola's dream of becoming a major commercial shipping port never materialized due to competition with neighboring seaports and a failure to modernize port facilities and diversify railroad services. Adding to Pensacola's economic woes was the failure of two local banks. The Pensacola State Bank folded on December 6, 1913[13] and the First National Bank failed on January 8, 1914.[14] Economic recovery was unlikely without an infusion of new commercial or government investments, both of which were doubtful in the immediate future.

NEW OPPORTUNITIES

Newspaper articles hopefully discussed a "new policy of specialization" by which closed Navy yards might be reopened in support of a specific operational field, although what special work might be done in Pensacola remained undetermined.[15] Nonetheless, the Pensacola Commercial Association continued to focus their efforts on the reopening of the Pensacola Navy Yard as an industrial shipyard. They were unaware that in the fall of 1913 the Secretary of the Navy had convened a special board to formulate comprehensive plans for the use of the Pensacola Navy Yard that involved experimentation with an entirely new technology.

FAVORABLE DEVELOPMENTS

An editorial in *The Pensacola Journal* noted prophetically that it would be a "… good thing for the country if a Democratic Administration soon gets control of things. Perhaps a new Secretary of the Navy can be found whose world is not measured by a few hundred miles."[16] Both of these wishes were soon granted

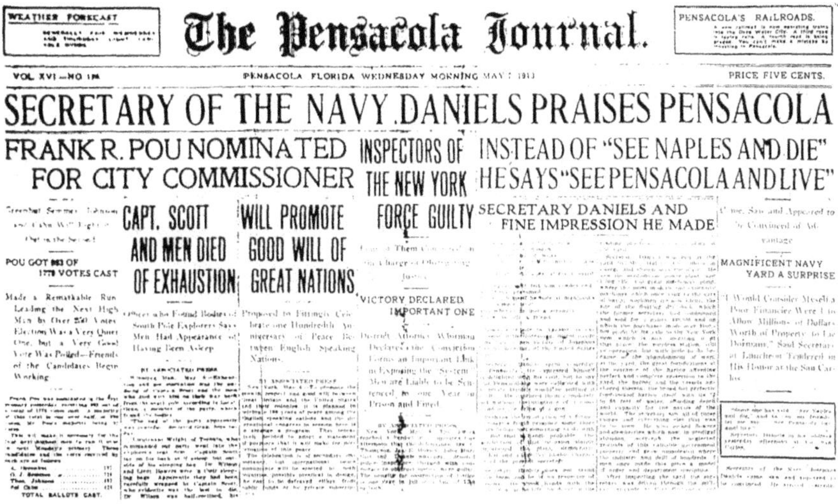

WEATHER FORECAST

The Pensacola Journal.

PENSACOLA'S RAILROADS.

VOL. XVI—NO. 1 PENSACOLA, FLORIDA, WEDNESDAY MORNING PRICE FIVE CENTS.

SECRETARY OF THE NAVY DANIELS PRAISES PENSACOLA

FRANK R. POU NOMINATED FOR CITY COMMISSIONER

POU GOT 963 OF 1779 VOTES CAST

INSPECTORS OF THE NEW YORK FORCE GUILTY

VICTORY DECLARED IMPORTANT ONE

INSTEAD OF "SEE NAPLES AND DIE" HE SAYS "SEE PENSACOLA AND LIVE"

SECRETARY DANIELS AND FINE IMPRESSION HE MADE

CAPT. SCOTT AND MEN DIED OF EXHAUSTION

WILL PROMOTE GOOD WILL OF GREAT NATIONS

MAGNIFICENT NAVY YARD A SURPRISE

Figure 2. In May 1913 Pensacola's efforts to reopen the Pensacola Navy Yard were rewarded by Secretary of the Navy Josephus Daniels' visit shortly after taking office. (Courtesy *Pensacola News Journal*)

when presidential elections brought the Democrats of Woodrow Wilson into power in March 1913 and Josephus Daniels, of North Carolina, was appointed the new Secretary of the Navy.

Almost immediately, Daniels announced that he would make a personal inspection of the southern navy yards.[17] His visit to Pensacola on May 6, 1913 was everything that Pensacolians hoped for, particularly his comment that "I would consider myself a poor financier were I to allow millions of dollars worth of property to lie dormant."[18] Yet while effusive in his praise of Pensacola, he also was non-committal on whether he would recommend reopening the yard. Nonetheless, Pensacola's lobbying efforts were further boosted when a Board of Survey of Shore Stations arrived on June 15, 1913 to conduct a thorough inspection of the yard.[19]

Shortly before their visit, a newspaper article revealed that Pensacola was being considered as the home for an "advance base outfit,"[a] a self-sustained expeditionary force of 2000 U.S. Marines equipped with three transports to make them capable of immediate response to any crisis.[20] Philadelphia, current home to the U.S. Marines and concerned over the potential loss of more than $2,000,000 a year in spending, attacked Pensacola in print, noting that:

Pensacola as a marine base in the estimation of the officers and men of the marine corps is a huge joke…the men who comprise the marine corps come from northern cities, and

[a] The imminent opening of the Panama Canal and the deepening crisis in Mexico certainly played a critical part in Pensacola's selection for this force.

they are not used to the climate and heat of the Florida station. Enlistments would be reduced and in time would mean the extinction of the corps.[21]

This prompted a quick counter by Mr. C.E. Dobson, President of the Pensacola Commercial Association:

If the marines are made of such poor stuff that they have to be carefully considered as to climate, then it would seem to be high time that they were given a special course in endurance, and if they are so constituted that it is necessary that they spend their leisure time in a city of many thousands, then it is time they were given a few lessons in discipline.[22]

Nonetheless, knowing full well that "...the future of the yard depends upon the report of this board" Pensacolians were encouraged by the fact that the old navy yard was being considered as a repair station, industrial shipyard and a U.S. Marine base. The Board's report released in September 1913 "...surpassed expectations of even the most ardent admirer of Pensacola and its advantages" and noted that "Pensacola is the best and only site practicable on the Gulf of Mexico for a Navy Yard of the first class."[23]

Events unfolded rapidly. On November 3, 1913 electric pumps at the water station at the yard were started although "No intimation of what the navy department intends doing has been received."[24] On November 17, 1913 Assistant Secretary of the Navy Franklin Delano Roosevelt visited for an inspection. He was reportedly enthusiastic about the yard and announced that "Approximately $65,000 in unexpended appropriations is now available for immediate use at the Pensacola Navy Yard in anticipation of its occupation as an advanced marine base."[25]

On November 25, 1913 the Secretary of the Navy ordered 750 U.S. Marines from Philadelphia to the Pensacola Navy Yard, the "...result of the favorable report of Assistant Secretary Roosevelt, who found the Pensacola reservation especially adapted to the advance base work being carried on by the Marine Corps."[26] The news article emphasized that "It is not intended to open the Pensacola yard for industrial purposes."[27] At this point Pensacolians were unsure whether anything beyond the stationing of U.S. Marines was planned, although it had been previously announced that there would be a "winter of naval activity in Pensacola" as several battleships, cruisers, torpedo boat destroyers and submarines were expected to visit after conducting winter maneuvers at Guantanamo Bay, Cuba.[28]

On November 26, 1913 Rear Admiral Homer Sanford, Chief of the Bureau of Yards and Docks, visited Pensacola for several days to prepare the Pensacola Yard for use as an advanced camp by the U.S. Marine Corps.[29] When asked by a reporter upon his departure "Is there any chance of the yard being opened as an industrial plant?" the Admiral replied "Well, you cannot tell what the future will bring forth...{in a}...non-committal answer."[30]

U.S. MARINES LAND

On December 2, 1913 the transport USS PRAIRIE with 800 Marines embarked under the command of Lieutenant Colonel John LeJeuene[b] arrived in Pensacola harbor.[31] On December 5, 1913 they debarked at the Pensacola Navy Yard and Lieutenant Colonel LeJeuene relieved Chief Carpenter Haley to take charge of both the Navy yard and the Naval reservation, unusual in that a Naval officer normally exercised this command.[32]

The Marines had their work cut out for them in making their living spaces habitable. The stripping of the Pensacola Navy Yard upon its closure had left the buildings derelict to the point where even the light fixtures "...were torn from all of the quarters and sent to Brooklyn where they were thrown into the scrap heap."[33] However, it was noted that even if the buildings were not ready for occupancy "...the marines are supposed to be equipped fully and can make themselves comfortable at any point."[34]

An immediate impact was felt on the local economy when 400 Marines received nearly $10,000 on their December 11, 1913 payday. *The Pensacola Journal* reported that "...nearly all of the men who were paid came up to the city and as a result of their visit, together with a number of bluejackets from the war vessels, the streets yesterday afternoon and last night had a lively appearance."[35] Another $10,000 was paid out to the remaining Marines the next day.

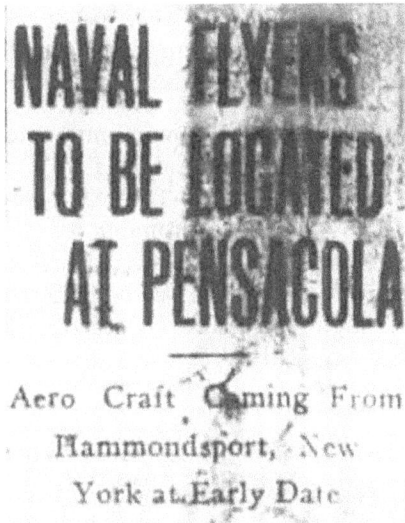

NAVAL FLYERS TO BE LOCATED AT PENSACOLA

Aero Craft Coming From Hammondsport, New York at Early Date

Figure 3. At the time of this December 1913 article, it was thought that the Naval aviators would stay in Pensacola only for the winter before returning to their summer aviation camp in Maryland. (Courtesy *Pensacola News Journal*)

NAVAL AVIATION ANNOUNCEMENT

On October 7, 1913 Secretary of the Navy Josephus Daniels, recognizing the somewhat disorganized aviation policy within the U.S. Navy, convened a board of officers to provide a "comprehensive plan for the organization of a Naval Aeronautic Service."[36] Of the many detailed recommendations made by this board in its report dated November 25, 1913, one "...recommended the selection of Pensacola as the place for the aviation centre and flying school."[37] In an incredible windfall for Pensacola, all of Naval aviation was being

---

[b] Later Major General and Thirteenth Commandant of the Marine Corps from July 1, 1920 to March 4, 1929. Of interest, in early 1913 LeJeune flew as a passenger with Naval aviators during winter maneuvers at Guantanamo Bay, Cuba.

transferred permanently here. Remarkably, very few, if any, Pensacolians knew this.

On December 18, 1913 a brief newspaper article titled "Naval Flyers to be Located at Pensacola" appeared on the front page of *The Pensacola Journal*.[38] The article further stated that "The naval flying machines, which have been undergoing tests at Hammondsport, N.Y. are expected to be sent to Pensacola for the winter, where the experiments will be continued."[39]

The press and public were under the impression that the aviators would spend only the winter in Pensacola before returning to their summer camp located near Annapolis, Maryland. This was not unusual, as the Naval aviators had left Maryland for California during winter of the previous year. A second newspaper article on January 2, 1914 also confirmed that Pensacola would serve as only the winter headquarters of the Naval Aeronautic Corps.[40]

On January 3, 1914 the monumental news finally broke. The front page of *The Pensacola Journal* heralded that "Pensacola is Designated As the Aeronautical Center of Navy."[41] The U.S. Navy's aviation training unit, and all of their aeroplanes and equipment, was to be permanently transferred to Pensacola. They were to establish the first Naval Aeronautic Station, a Naval Flying School, and use the USS MISSISSIPPI (BB-23) as an experimental aviation ship for flight operations. Pensacola was about to earn her title as "The Cradle of Naval Aviation."

Journal.

THE YEAR 1914

Is Going to Be the Best in Pensacola's History. We're all going to make it so.

JANUARY 3, 1914.   PRICE FIVE CENTS.

Pensacola is Designated As the Aeronautical Center of Navy

Figure 4. On January 3, 1914 the first indication for Pensacolians that Naval aviation had found a permanent home in Pensacola was this headline. (Courtesy *Pensacola News Journal*)

While this was exciting news, Pensacolians were still unduly focused on the reopening of the yard as an industrial shipyard to boost the local economy. No one could imagine that within a few short years the economic impact of Naval aviation in Pensacola would be measured in millions of dollars, far exceeding the expenditures of the Pensacola Navy Yard.

Figure 5. Photograph of NAS Pensacola's waterfront in 1914, illustrating the daunting task Navel aviators faced in overcoming years of neglect to convert an abandoned shipyard into the world's finest aeronautical station. Note the 10 tent hangars built on the flying beach for the aviators' fragile seaplanes and hydroaeroplanes. (Courtesy National Museum of Naval Aviation)

NAVAL AVIATION LANDS

The assignment of the USS MISSISSIPPI as mothership to the fledgling aviators was a good omen. Named after the nearby "Father of Waters," one of Old Miss' first port calls after her commissioning on February 1, 1908 was Pensacola. The collier USS ORION (AC-11) was also assigned to assist in the transportation of the aviation unit.

Embarked aboard both vessels was a 32-man unit of flight instructors, students, mechanics, sailors, aviation equipment and supplies. Civilian technical representatives from each of the three aeroplane manufacturers represented in the unit's complement of hydroaeroplanes and seaplanes also accompanied them. Lieutenant Commander Henry Mustin, Naval Aviator Number 11, commanded both the Naval Aeronautic Station and the USS MISSISSIPPI while Lieutenant John Towers[c], Naval Aviator Number 3, commanded the Naval Flying School. They faced a daunting task.

The USS ORION arrived at Pensacola on the evening of January 19, 1914. The next day the USS MISSISSIPPI took "...berth off the navy yard, dropped her big anchors where the big ship remained for the day, and where, it was stated, will remain until the program for the aeronautical tactics is worked out."[42] Shortly thereafter, the USS MISSISSIPPI moved alongside the Pensacola Navy Yard's dock to provide power ashore. She also provided quarters for the aviation detachment since there were no suitable accommodations ashore.

AERONAUTIC STATION PREPARATIONS

Years of neglect and violent hurricane damage had left the Pensacola Navy Yard's buildings derelict and her once pristine grounds and beaches strewn with debris. Although the Marines had been ashore for more than a

Figure 6. Early 1914 sea level view of Naval Aeronautic Station, Pensacola tent hangars and aeroplanes along their flying beach. (Courtesy National Museum of Naval Aviation)

[c] The first career aviator to be promoted to the rank of full admiral.

month, they were still struggling with their own crude accommodations. As Lieutenant Commander Mustin and Lieutenant Towers stepped ashore to survey the damage, little could they envision that within a few short years this small southern outpost would soon develop into the largest Naval Air Station in the world.

Lieutenant Pat Bellinger, Naval Aviator Number 8, described his first impressions:

The Yard was overgrown with shoulder high grass and weeds, all docks except one were partly washed away and along a nice sandy beach were a lot of piles of driftwood and rocks that had to be removed. I could see there was a lot of work to be done; however, the weather was fine and I had to admit it was salubrious.[43]

In Lieutenant Commander Mustin's first report to Washington, dated the day of his arrival, he noted:

Found some buildings quite usable, a construction and repair shop full of excellent tools, a lookout tower; however, the beach was in a fearful state with wreckage of all kinds, bricks, stones and old railroad iron. This is the reason it will take two weeks to start the school. Tomorrow will start ship's crew clearing away wreckage and build-ing runways. This morning it looked like the ruins of a prehistoric city.[44]

The establishment of a flying beach, critical to the launching and recovery of the U.S. Navy's flying boats and hydroaeroplanes, was the highest priority for the aviators. Placing an advertisement in the newspaper, they sought "…to open negotiations with some parties in Pensacola for the clearing of fifteen acres on the naval reservation for the use of the aeroplanes…{which would be} …used to start from and to alight upon when the airships are performing their aerial feats."[45] Dynamite was also needed to clear the flying beach of obstructions.

① SCOFIELD ② CORRY ③ BRONSON ④ HAAS ⑤ PAUNACK ⑥ EVANS ⑦ JOHNSON ⑧ READ ⑨ MUSTIN ⑩ BELINGER ⑪ CUNNINGHAM ⑫ SAUFLEY ⑬ NORFLEET ⑭ EDWARDS ⑮ BARTLETT ⑯ SPENCER ⑰ McDONNELL.

Figure 7. Some of Pensacola's first Navy and Marine Corps aviators stand beside a Curtiss Flying Boat in this 1915 photograph.. Several airfields in the Pensacola area were named after some of these early aviation pioneers. (Courtesy U.S. Navy)

Although weeks of strenuous effort would be required before the flying beach[d] would be considered suitable for flight operations, the Naval aviators immediately began to offload their disassembled Curtiss flying boats, Curtiss hydroaeroplane, Wright hydro-aeroplane and Burgess flying boat.

To protect these fragile craft, white canvas tent hangars with floors made of cinders were erected along the sandy flying beach.[46] Heavy timbers found on the shore were used to construct a seawall in front of the hangars and "…carried the jackstays for the front flaps, making the hangars better able to withstand severe windstorms."[47] Wooden runways were also constructed from each hangar into the water to a depth a three feet, as the sand was too soft to maneuver the aeroplanes.

Although initially equipped with only a handful of aeroplanes, future plans published a few months later called for the arrival of dirigibles, dozens of additional aeroplanes and the construction of dirigible sheds and permanent wooden hangars to replace the temporary tents.[48]

OFFICIAL DESIGNATION

In January 1914 the Pensacola Navy Yard was reopened as a Naval Aeronautic Station under the command of Lieutenant Commander Mustin aboard the USS MISSISSIPPI. An administrative reorganization on November 16, 1914 shifted command to headquarters ashore and the official designation was Naval Aeronautic Station, Pensacola, Florida.

The official birth date of NAS Pensacola is considered February 11, 1914 when the first entry in the station logbook was made. An unknown author wrote:

J. GOSLING

The Secretary of the Navy has decided that the science of aerial navigation has reached that point where aircraft must form a large part of our naval force for offensive and defensive operations. Nearly all countries which have a navy are giving attention to this subject. This country has not fully recognized the value of aeronautics in preparation for war, but it is believed that we should take our proper place.

In December 1917 the station's title was administratively changed to Naval Air Station, Pensacola, Florida, as it is still known today.

NAS Pensacola's official insignia is that of "J. Gosling," created by Mr. Edward J. Collins, a draftsman in the Overhaul and Repair Department, in 1931. Predating the creation of Donald Duck by nearly three years, a young duck was selected to represent a flight student's "certain degree of foolishness and amusement" while learning to fly.

---

[d] Located between the old Pensacola Navy Yard's central wharf and the southwest coal wharf.

Figure 8. The Coast Guard's first pilot, Lieutenant Elmer Stone shown above in Pensacola in 1917, served as a co-pilot aboard the NC-4, the first aeroplane to cross the Atlantic Ocean. (Courtesy National Museum of Naval Aviation)

Despite all this proof to the contrary, Pensacolians were apparently still a little doubtful that the Naval aviators were here to stay. However, *The Pensacola Journal* reassured them that "The fact the navy department is contemplating these large expenditures here to prepare the Pensacola station for experimental work in aviation seems to offer conclusive proof that the department intends that the aeronautical school at Pensacola be a permanent institution."[49]

U.S. MARINE CORPS AVIATION LANDS

If Naval aviation was small, then Marine Corps aviation was nearly microscopic. The first U.S. Marine Corps aviator, First Lieutenant Alfred A. Cunningham, had been relieved of flight duty at his own request on August 11, 1913. This left only First Lieutenant Bernard L. Smith, Marine Corps Aviator Number 2 and Naval Aviator Number 6, and Second Lieutenant William M. McIlvain, Marine Corps Aviator Number 3 and Naval Aviator Number 12, on flight duty.

While the Naval aviation camp was relocating to Pensacola in January 1914, First Lieutenant Smith and Second Lieutenant McIlvain established a Marine aviation element at Culebra, Puerto Rico. Utilizing a Curtiss flying boat and hydroaeroplane, they conducted combined operations with the Marine Advance Base Brigade. Upon the conclusion of the exercise, the aviation detachment was disbanded and the aviators embarked with 1,700 other Marines aboard the USS HANCOCK and USS PRAIRIE to rejoin their Navy

---

[e] I have found several references to the presence of a "Marine Corps Flying Station" in Pensacola before the arrival of Naval aviation in January 1914 but the historical record does not support this. There was a U.S. Marine Corps flying facility at the Philadelphia Navy Yard in late 1913, which may explain the confusion.

counterparts at Pensacola.[e] They arrived at Pensacola on February 15, 1914 to form the 'Marine Section' of the Naval Flying School.[50] It was not until World War I that Marine aviation separated from Naval aviation to form their own independent units.[51]

U. S. COAST GUARD AVIATION LANDS

On April 1, 1916 Second Lieutenant Elmer F. Stone, United States Coast Guard, later joined by Second Lieutenant Charles E. Sugden, United States Coast Guard, reported to Pensacola for flight training. Lieutenant Stone successfully completed flight training on April 10, 1917 and was designated U.S. Coast Guard Aviator Number One and Naval Aviator Number 38.

To this day, officers from the U.S. Navy, U.S. Marine Corps and U.S. Coast Guard train together at Pensacola to earn their wings of gold. Officers from several countries and the U.S. Air Force also join them.

OPEN HOUSE AT NAVAL AERONAUTIC STATION, PENSACOLA, FLORIDA

Preparations for the Naval aviators' first flight at Pensacola continued at a breakneck speed. On January 24, 1914, an article appeared in *The Pensacola Journal* that read:

The aviators at the aeronautical school on the naval reservation and officers of the battleship Mississippi have announced that when the aeroplanes and hydroplanes are set up and all is in readiness for the aerial practices and instructions, openhouse will be the order at the navy yard. The public is invited to attend and see the flying, and one of the officers said he would be glad to take any Pensacolian along as a passenger in his aerial journeys. Ladies as well as gentlemen will be given an opportunity to fly with Uncle Sam's airmen.[52]

The first flight by the United States Navy above Pensacola's azure bays and clear water bayous was less than nine days away.

NOTES

[1] Meyer Seals Doom of Pensacola Yard. (1911, June 29). *The Pensacola Journal.*

[2] Stores at Pensacola Yard Ordered Packed. (1911, July 4). *The Pensacola Journal.*

[3] The Sterling Leaves With a Large Cargo. (1911, October 19). *The Pensacola Journal.*

[4] Local Navy Yard Officially Closes. (1911, October 21). *The Pensacola Journal.*

[5] Is Inspecting the Wireless Station Here. (1913, August 1). *The Pensacola Journal.*

[6] Obeying Meyer's Orders. (1911, November 25). *The Pensacola Journal.*

[7] Obeying Meyer's Orders. (1911, November 25). *The Pensacola Journal.*

[8] Moving Navy Yard Supplies to Charleston. (1912, April 17). *The Pensacola Journal.*

[9] Moving Navy Yard Supplies to Charleston. (1912, April 17). *The Pensacola Journal.*

[10] Pensacola Yard Now Dark. (1912, May 26). *The Pensacola Journal.*

[11] Want Former Policy of Navy Again Adopted. (1912, June 11). *The Pensacola Journal.*

[12] McGovern, J.R. (1976). *The Emergence of a City in the Modern South: Pensacola 1900-1945.* DeLeon Springs, FL: E.O. Painter Printer Company.

[13] Pensacola State Bank Stockholders Meet. (1913, December 6). *The Pensacola Journal.*

[14] First National Bank Suspends London Failure a Partial Cause. (1914, January 8). *The Pensacola Journal.*

[15] To Place Navy Yards in South in New Service. (1911, November 15). *The Pensacola Journal.*

[16] The Secretary and the Navy. (1911, November 4). *The Pensacola Journal.*

[17] Secretary of the Navy Says He Will Visit The Pensacola Yard Soon. (1913, April 1). *The Pensacola Journal.*

[18] Secretary of the Navy Daniels Praises Pensacola. (1913, May 7).*The Pensacola Journal.*

[19] Admiral Edwards and Members of Survey Board in Pensacola to Look at Navy Yard. (1913, June 16). *The Pensacola Journal.*

[20] Pensacola as Important Naval Base. (1913, June 9). *The Pensacola Journal.*

[21] Asks Bids For Moving 1,000 Marines to Pensacola. (1913, July 6). *The Pensacola Journal.*

[22] The Philadelphia Yard and Pensacola Weather. (1913, July 17). *The Pensacola Journal.*

[23] First Class Navy Yard for Pensacola. (1913, September 3). *The Pensacola Journal.*

[24] Pumps Start at Yard. (1913, November 4). *The Pensacola Journal.*

[25] $65,000 Available for Use at Pensacola Navy Yard. (1913, November 18). *The Pensacola Journal.*

[26] Transport Prairie Leaves Philadelphia Soon with 750 Marines for Pensacola Yard. (1913, November 25). *The Pensacola Journal*

[27] Transport Prairie Leaves Philadelphia Soon with 750 Marines for Pensacola Yard. (1913, November 25). *The Pensacola Journal.*

[28] Winter of Naval Activity. (1913, November 10). *The Pensacola Journal.*

[29] Prepare Pensacola Navy Yard for Reception of Marines. (1913, November 27). *The Pensacola Journal.*

[30] Admiral Will Return Today to Washington. (1913, December 3). *The Pensacola Journal.*

[31] Transport Prairie Arrives in Port; Marines Debark Today. (1913, December 3). *The Pensacola Journal.*

[32] Marines will Debark Today from Prairie. (1913, December 5). *The Pensacola Journal.*

[33] Will Supply Electricity for Navy Yard. (1913, December 10). *The Pensacola Journal.*

[34] Transport Prairie Arrives in Port; Marines Debark Today. (1913, December 3). *The Pensacola Journal.*

[35] Pay Day With the Marines at Pensacola. (1913, December 12). *The Pensacola Journal.*

[36] Turnbull, A.D. & Lord, C.L. *(1949) History of United States Naval Aviation.* New Haven, CT: Yale University Press.

[37] Daniels Defends His Course In Assisting Southern Yards. (1915, April 12). *The Pensacola Journal.* Note: provides reasons Pensacola was selected for aeronautical station.

[38] Navy Flyers to be Located at Pensacola. (1913, December 18). *The Pensacola Journal.*

[39] Navy Flyers To be Located at Pensacola. (1913, December 18). *The Pensacola Journal.*

[40] Battleship Mississippi Will Bring Aeronautic Corps Here. (1914, January 2). *The Pensacola Journal.*

[41] Pensacola is Designated As the Aeronautical Center of the Navy. (1914, January 3). *The Pensacola Journal.*

[42] Mississippi is Here to Make Lenghty (sic) Stay. (1914, January 21). *The Pensacola Journal.*

[43] Bellinger, P. (n.d.). *The Gooney Bird.* Unpublished Manuscript. Washington, D.C.: Naval Historical Center, Operational Archives Branch. NRS 1982-2.

[44] Pensacola Celebrates Fiftieth Year. (1964 January). *Naval Aviation News,* pages 8-9.

[45] Aviators Are to Clean Off Fifteen Acres. (1914, January 22). *The Pensacola Journal.*

[46] Birth Of An Air Station. (1938, July). *The Gosport.*

[47] Birth Of An Air Station. (1938, July). *The Gosport.*

[48] Bright Outlook For The Pensacola Yard. (1914, May 5). *The Pensacola Journal.*

[49] Aviators Are To Clean Off Fifteen Acres. (1914, January 22). *The Pensacola Journal.*

[50] Transports Hancock and Prairie Bring 1,700 Men. (1914, February 16). *The Pensacola Journal.*

[51] Johnson, E.C. (1977). *Marine Corps Aviation: The Early Years 1912-1940.* Washington, DC: U.S. Marine Corps.

[52] Local People Invited To Go Up In Airships. (1914, January 24). *The Pensacola Journal.*

CHAPTER 6
FIRST NAVAL AVIATION FLIGHT OVER PENSACOLA

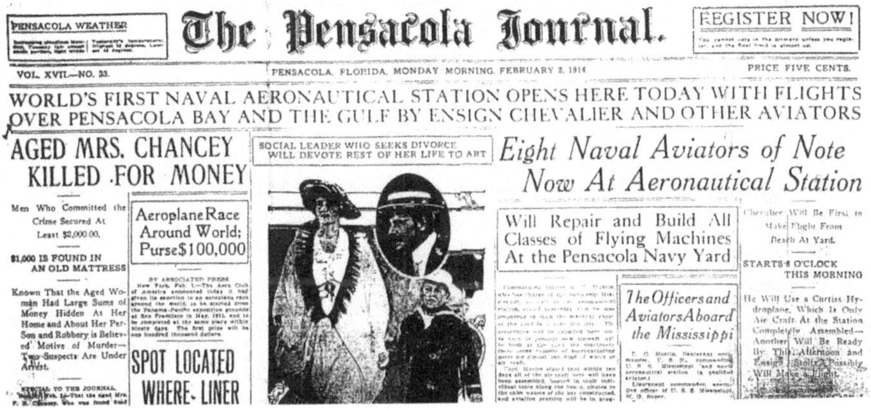

Figure 1. Only 13 days after arriving in Pensacola, Naval aviators were ready for their first flights. The newspaper the next day stated that "the machines in the air look like gigantic buzzards and make a noise plainly audible on the ground as the airmen pass over." (Courtesy *Pensacola News Journal*)

FIRST NAVAL AVIATION FLIGHT AT PENSACOLA

On a brisk winter morning nearly three years after Nels Nelson's record-setting flight, a banner headline in *The Pensacola Journal* announced that the "World's First Naval Aeronautical Station Opens Here Today With Flights Over Pensacola Bay And The Gulf By Ensign Chevalier And Other Aviators."[1] The date was February 2, 1914 and Pensacola weather called for ideal flying conditions: light winds with morning temperatures in the 40s and afternoon temperatures in the 60s.

At the Naval Aeronautic Station, the aviators were busy. From a canvas tent hangar, beach crews pushed a wooden hulled Curtiss F-Boat designated the C-5 down a ramp into the frigid waters of Pensacola Bay. Scrambling aboard were Lieutenant John Towers, Naval Aviator Number 3 and Commander of the Naval Flying School, and flight student Ensign Godfrey de C. Chevalier, future Naval Aviator Number 7. After completing a series of engine and control checks, they gently maneuvered their frail flying boat out into Pensacola Bay and

Figure 2. Barely visible in this photograph of the first Naval aviation flight in Pensacola's history is Lieutenant John Towers in the left seat and Ensign Chevalier in the right. (Courtesy National Museum of Naval Aviation)

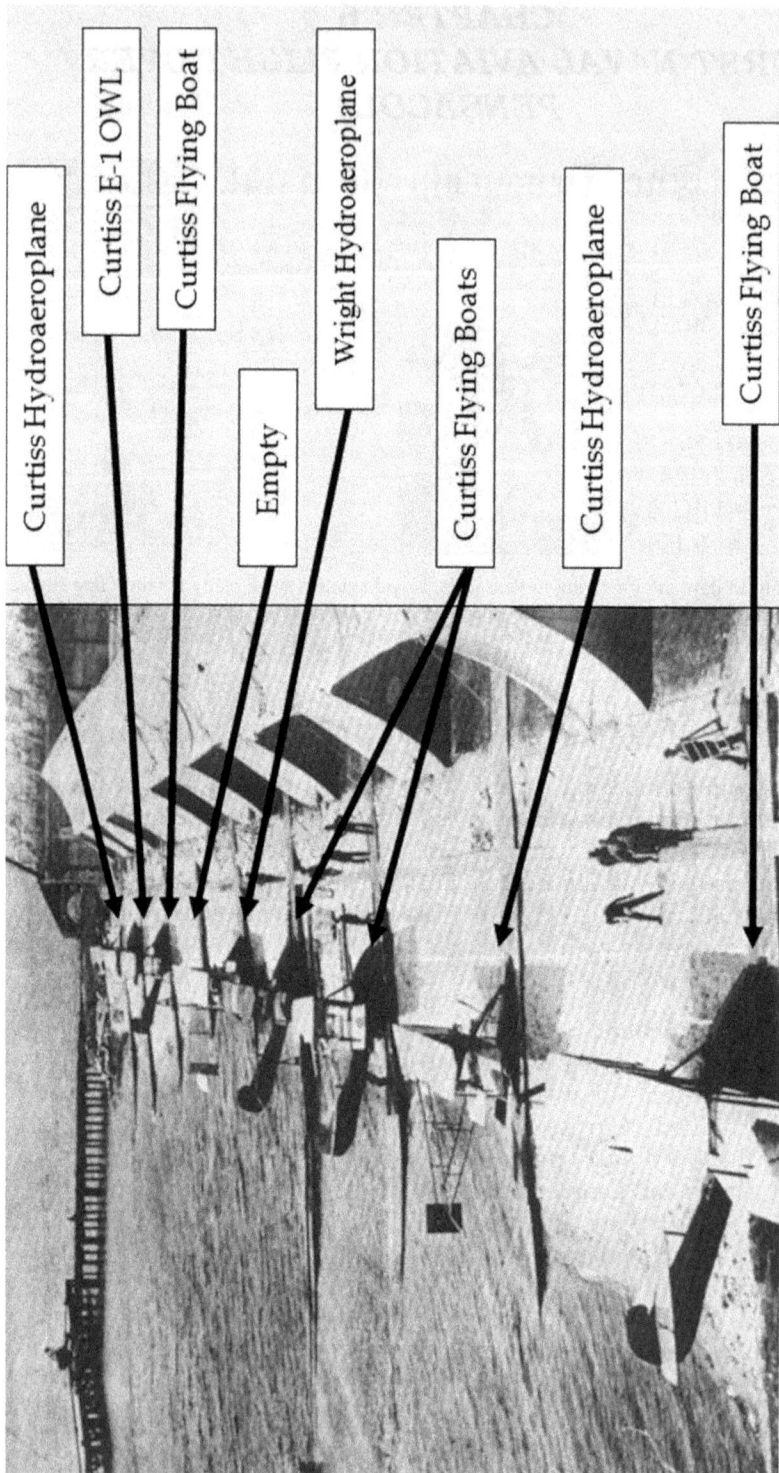

Figure 3. Naval aviation's first aeroplanes in Pensacola in front of their canvas hangars at the aeronautic station's flying beach. Seven aeroplanes arrived on January 20, 1914 with the Naval aviators (Burgess-Dunne D-1, Curtiss hydroaeroplane A-3, Wright hydroaeroplane B-3, and Curtiss flying boats C-1, C-2, C-4, and C-5). Two additional aeroplanes arrived on February 16, 1914 with the Marine aviators (Curtiss hydroaeroplane E-1 OWL and Curtiss flying boat C-3). (Courtesy National Museum of Naval Aviation)

carefully estimated the wind's direction while keeping a watchful eye on the size of the wave swells. After sighting down their flight path to clear for possible obstacles, Lieutenant Towers firmly applied the throttles, opening a deafening roar from the engine and propeller spinning only inches behind his head. Water sprayed through the propeller and over the gunwales as the flying boat slowly rose on step, finally breaking the bay's icy grip at about 9:30 a.m.. Choosing to stay safely close to the station in the event of a malfunction, the aviators commenced a leisurely climb to a height between 800 and 1,000 feet, flying for 20 minutes before passing over Bayou Grande and alighting gently back on the bay. The Cradle of Aviation was baptized.

The next day *The Pensacola Journal* described the Naval aviators' first flights from Pensacola:

MANY FLIGHTS MADE MONDAY BY NAVY MEN

Ensigns of Local Flying School, Who are Listed Among World's Most Successful Aviators, Found Local Atmospheric Conditions Ideal for Aeronautics.

Three flights were made by aviators of the government naval aeronautical school at the navy yard yesterday. A Curtiss flying boat, No. C-5 was used, being the only one of the eight aircraft of the local station which was completely assembled.

Lieutenant Jack Towers, who is in charge of the aviation camp, and Ensign Godfrey de C. Chevalier made the first flight, reaching an altitude of between eight hundred and one thousand feet. They left the runways at about nine-thirty yesterday and were out about twenty minutes, flying over Bayou Grande and Pensacola bay.

Figure 4. A much clearer photograph of the Curtiss Flying Boat C-5 on beaching gear. This was the first U.S. Navy aeroplane to fly over Pensacola. It sits next to building 38, which is still in use today. (Courtesy Pensacola Historical Society)

Ensign Chevalier with Lieutenant Commander Roper as a passenger, made a second flight, remaining out about fifteen minutes. This was the first aerial voyage for Lieut. Roper, who is being taught to fly. The greatest altitude for this trip was five hundred feet. The flying boats are equipped with instruments to show at all times in what altitude the airship is.

Lieutenant Towers, with Captain (sic) Mustin of the Mississippi, who is also a man-bird, made the third flight. These navigators of the air flew about at an altitude of eight hundred feet for some time.

A number of flights will be made today.

Yesterday's flying was concluded about eleven-thirty o'clock. All of the rises and landings, which are made from and on the water respectively, were perfectly executed. The atmospheric conditions were ideal for aeronautics yesterday. At one time the aviators came over a point near the western portion of the city and disappeared to the northward.

The machines in the air look like gigantic buzzards and make a noise plainly audible on the ground as the airmen pass over. The speed attained is remarkable, being much greater than that of a buzzard or hawk in flight. In descending the aviators experience, until they become used to flying, a feeling of nausea like unto that feeling one has when descending rapidly in an aviator (sic).[2]

The ice broken with their first flight, Pensacola's Naval aviators reassembled and started flying their other aeroplanes to the thrill and delight of Pensacolians who had a ringside seat to aeronautical displays enjoyed by few people in the world. However, the jubilation of these first flights would be tempered by a tragic event two weeks later.

NOTES

[1] World's First Naval Aeronautical Station Opens Here Today with Flights Over Pensacola Bay and the Gulf by Ensign Chevalier and Other Aviators (1914, February 2). *The Pensacola Journal.*

[2] Many Flights Made Monday By Navy Men. (1914, February 3). *The Pensacola Journal.*

CHAPTER 7
FIRST AVIATION DEATHS IN PENSACOLA

Figure 1. The Pensacola Journal headline the day after Lieutenant (Junior Grade) Murray became the first aviator to be killed in Pensacola. (Courtesy *Pensacola News Journal*)

FIRST AVIATION FATALITY AT PENSACOLA

On February 16, 1914 Lieutenant (Junior Grade) James Murray, flying the D-1 Burgess flying boat over Pensacola Bay, stalled, dived and crashed to his death from an altitude of 200 feet. This was the first aviation fatality in the history of the state of Florida and the second Naval aviator fatality in their short history. *The Pensacola Journal* described the fatal accident:

LIEUT. MURRAY, OF AVIATION CORPS, DASHES TO DEATH. LOSES CONTROL OF MACHINE AND FALLS 200 FEET. CAUGHT BENEATH WRECKAGE OF HYDRO AEROPLANE IS DROWNED.

A Normal Glide Suddenly Turn its Nose Downward and Strikes the Water With Terrific Force

Lieut. J.M. Murray, U.S.N., attached to the Pensacola aeronautical station, met death yesterday afternoon at 4 o'clock when making a flight over Pensacola Bay. Hundreds of horrified spectators gathered along the waterfront or viewing the flight from office buildings in the city, saw the hydroaeroplane, or flying boat, take a fatal dip when at

Figure 2. Lieutenant (Junior Grade) Murray was piloting this Burgess D-1 flying boat when he was killed. (Courtesy Aerofiles)

a height of about 200 feet and strike the water with terrific force. Lieut. Murray, it is believed, was drowned by being caught beneath the wreckage. His body arose to the surface when a motor boat towed the wrecked machine from where it had fallen. The remains were quickly conveyed to the Battleship Mississippi after Lieuts. Saufley and Bellinger had applied all first aid methods without success. The cause of the accident has not yet been determined. Either some portion of the machine broke while the aviator was gliding downward or else he lost control of it through some other force of circumstances and was unable to get control of the machine. The latter its fatal dip.(sic)

THIRD TRIP OF THE DAY

The flight was the third one of the day for Lieut. Murray. He had made two very successful flights during the forenoon and went out a few minutes before 4 o'clock for another flight. He was seen by hundreds of people as he passed this city, went up the bay for a considerable distance, crossed over towards Town Point, made a circle and started back towards the city. He was then at a much greater height than at any time during his flight of about 15 minutes duration, and started a glide. He had come down for a considerable distance when the machine, taking a normal glide, was seen to turn its nose downwards, and in a moment, it was evident to many of the spectators that the aviator had lost control of the machine. The latter struck the water with much force and was wrecked.

ACCIDENT REPORTED

The accident was witnessed by the lookout of the battleship Mississippi who immediately reported, and Lieutenants Bellinger and Saufley put out in a flying boat, while Weather Observer Reed, who was also following the flights from his office window, phoned to several motor boats about the city. Private Murphy on the mine planter Gen. Schofield, notified Lieut. Peek, commander of the vessel, and the latter hurriedly put off in a small boat to the scene. The aviators in the flying boat were the first to reach the scene of the wreck and they were unable to find any trace of the body. Lieut. Bellinger left the flying boat and climbed aboard the wreck and endeavored to get some assistance from the motor boats which had gone out from the city, but none of them would come near him. Finally Lieut Peek arrived in his boat and took the wreckage in tow. As soon as it was moved the body of Lieut. Murray rose to the surface. It had been held beneath the water by the wreckage and it is possible that if assistance had arrived and the wreck moved soon enough his life could have been saved.[1]

MORE NAVAL AVIATOR DEATHS AT PENSACOLA

Nearly 15 months passed before the next fatality occurred in Pensacola. On May 8, 1915 Lieutenant Marvin L. Stolz was killed when "...the hydro-aeroplane in which he was making a flight, took a sudden dive downward, falling about one hundred feet, striking in the bay. The radiator of the machine struck the aviator at the base of the skull and this, it was stated, had caused his death."[2]

As we will discuss later, the cause of Lieutenant Stolz's death is significant and the continued use of pusher aeroplanes will lead to a minor mutiny by the Pensacola Naval aviators.

The third aviation fatality in Pensacola occurred on May 24, 1916 when Lieutenant J.V. Rockwell was killed while flying directly in front of the aeronautic station.[a] An aileron snapped on his aeroplane and the "…machine suddenly made a 'nose dive' – a perpendicular plunge – to the water, and the watchers on shore knew from the gyrating motion of the plane that he had lost control."[3] Lieutenant Rockwell left behind a wife and three children, with the oldest child only seven years old. Fortunately, only a few months before, a new regulation had gone into effect that provided the widow of an aviator double pension in addition to a full year's salary, which to Mrs. Rockwell amounted to $60 a month.[4]

FIRST NAVAL AVIATION FATALITY

The first death in Naval aviation history occurred on June 20, 1913 near Annapolis, Maryland. A sudden gust of wind threw Ensign William Billingsley from the Wright B-2 hydroaeroplane he was piloting and he fell 1,500 feet to his death in the water.

His co-pilot was Lieutenant John Towers, who later assumed command of the first Naval Flying School at Pensacola. He also was thrown from his seat but managed to grab an upright and hold on until the B-2 crashed. Although severely hurt, Lieutenant Towers recovered from his injuries and became a firm believer in the requirement for shoulder straps to secure pilots in their seats.

This fatal accident occurred nearly three and a half years after the very first Naval officer, Lieutenant Theodore Ellyson, was ordered to report for flight training in December 1909 at the Glenn Curtiss Aviation Camp at North Island, California.

FURTHER DEATHS

As the number of flight students increased at Pensacola so too did the number of fatal accidents. From February 1914 to the end of World War I on November 11, 1918, 25 young men were killed and 27 wounded as a result of aeroplane crashes.[5]

Pensacolians witnessed many of these accidents, including spectacular mid-air collisions and burning or stalled aeroplanes hurling themselves into the ground or the bay. All too often Pensacola's cemeteries embraced the broken bodies of her young adopted aviators as they paid for their flying mistakes with their lives or pushed the operating envelope of their frail bamboo and fabric flying machines too far.

---

[a] A civil engineer and the station's planning superintendent when he volunteered for flight training, some aviation history texts erroneously identify Lieutenant Rockwell as the first civilian aviation casualty in Pensacola.

One of the saddest deaths was that of Lieutenant R.C. Saufley, the fourth Naval aviator to die at Pensacola. He was flying over Santa Rosa Island on June 10, 1916 attempting to break his own record for sustained flight when his hydroaeroplane suddenly spun into the island.[6] Lieutenant Saufley was the current world's record holder for altitude, at 16,072 feet, and endurance, eight hours and five minutes. He had been airborne for eight hours and fifty-one minutes, breaking the endurance record, when he crashed.

PUSHER VERSUS TRACTOR AEROPLANES

Striving to fly faster, higher and longer, each death, no matter how tragic, served in some way to advance these aviation pioneers' knowledge of aerodynamics, aeroplane construction, aeroplane performance, and piloting skills. While pilot error emerged as the leading cause of aeroplane accidents, a controversy developed on the placement of engines and propellers in relation to the cockpit.

In the "pusher" type aeroplane of this era, the propeller and engine were located in the rear of the flying machine. The aviator sat forward to balance the weight of this aft engine. This was the only type of aeroplanes the early Naval aviators had in Pensacola.

In the "tractor" type aeroplane the propeller and engine were located in the front of the cockpit. Glenn Curtiss had designed and flown a tractor as early as 1911, but the lack of an exhaust pipe, oil spray, poor visibility, and propeller blast had convinced Curtiss to favor the pusher design for his aeroplanes.[7]

The significance of engine placement drew considerable aviator concern when the deaths of Stolz, Rockwell and Saufley were attributed to the pusher aeroplane design. Lieutenant Pat Bellinger,[b] a Naval aviator in Pensacola at the time, described the issue best:

...as a result of flying the pusher type of plane, the majority of crashes resulted in fatalities. I, with many others, worked on many of those who were rescued unconscious from a crash. They all must have drowned or died quickly from their impact with the water and the motor falling on them. They all had severe wounds in the back of their heads. To try to prevent this, a padded wooden back was secured to the seats and extended up behind the head. This helped, but did not prevent fatalities.[8]

Pensacola Naval aviators recommended that tractor planes be purchased to replace the pushers, but Captain Mark Bristol, the current Director of Naval Aeronautics, rejected this[c]. After another student's death was attributed to the pusher design a telegram was sent to Washington drawing a line in the sand. The message stated: "As a result of conference of all aviators at Pensacola, unless otherwise directed there will be no further flying of pusher type of planes pending purchase of tractor planes previously recommended."[9]

---

[b] Later a Vice Admiral and Commander, Air Force Atlantic.

[c] In December 1913 Captain Bristol relieved Captain Washington I. Chambers, who is considered to be the "Father of Naval Aviation" for his pioneering efforts.

Washington was furious but the desired point had been made. As noted by Bellinger their goal was achieved when "As a result of the Pensacola dispatch, we were informed that tractor planes would be purchased, but we didn't get what we expected until the N-9s arrived in 1917. Anyway, flying was resumed with the pushers very shortly as we thought we had won our point."[10]

Eventually the Chief of Naval Operations, Admiral William S. Benson, ordered that all the old pusher hydroaeroplanes be grounded and replaced with the tractor type.[11]

In surplus sales after World War I, civilian aviators around the country purchased pushers in great numbers to establish some of this nation's first commercial aviation ventures. As will be seen later, former Naval aviators in Pensacola did the same to establish Pensacola's first commercial aviation enterprises.

Figure 3.   Hand launching an N-9 at NAS Pensacola. (Courtesy National Museum of Naval Aviation)

PENSACOLA RECEIVES TRACTORS

The first tractor aeroplane to arrive at Pensacola was a Martin Model S seaplane, followed by a rapid succession of Thomas-Morse, Sturtevant, and Curtiss designs. Of them all, the Curtiss N-9, a seaplane based upon the U.S. Army's successful JN-4B "Jenny" landplane, proved to be the most successful. The first N-9 arrived in Pensacola in late 1916 and nearly 600 N-9s eventually were purchased by the U.S. Navy.[12]   It was the first U.S. Navy aeroplane with installed seat belts and the first with Deperdussin controls.[13] The N-9 trained prospective Naval aviators in Pensacola for more than a decade,[d] and nearly all of the Naval aviators trained during World War I received training in this seaplane. [14]

Figure 4. Many hands make light work of hauling this seabird up one of NAS Pensacola's many seaplane ramps. (Courtesy U.S. Navy)

---

[d] The N3N "Yellow Peril" aeroplane so fondly associated with Naval aviation training did not arrive in Pensacola until August 1936.

For those who covet the few remaining N-9s preserved in museums, the next paragraph, published in *The Pensacola Journal* in December 1927, may be somewhat difficult to read:

The last of the N-9 type training planes have passed from the picture of naval aviation. These training seaplanes, of a type brought out during the war, gave wonderful service for nearly ten years and were forced out only by the strides of aviation and the great improvement in modern plane designing. Their place has been taken by the NY-2 type of training seaplane.

Figure 5. Like so many horses tied to a saloon's hitching posts, these N-9s resting along the Warrington flying beach patiently await their next rider. (Courtesy National Museum of Naval Aviation)

Over sixty of these N-9s have been placed on the scrap heap and burned. The sight of these planes going up in flames brought somber faces to many of the navy fliers, who had an attachment for this type plane. Perhaps the reason so many of the pilots liked the old N-9 was because of the fact that the planes could withstand crashes and bad smash-ups without serious injury to the occupants. There have been many N-9 crashes where the plane was reduced to match sticks and nobody was more than scratched. It is to be hoped that the NY-2s will even surpass this record, but the inherent stability of this type plane is so much greater than that of the N-9s that there are certain to be far less crashes. [15]

WRIGHT CAMP VERSUS CURTISS CAMP

A second controversy that developed involved the difference in performance between the aeroplanes manufactured by the Wright and Curtiss companies. Both the U.S. Navy and U.S. Army owned Wright and Curtiss aeroplanes and aviators from both services had received flight training at each company's facilities. This created competitive cadres of "Wright Men" and "Curtiss Men," who felt very strongly about the merits of their respective company's aeroplanes.

As the flight controls of the Wright and Curtiss aeroplanes differed significantly, aviators could not easily transition between the two manufacturers' aeroplanes without very specific training.

One early aviator noted:

A difficult situation had always existed in aviation, due to lack of universal methods of aircraft control. The Wrights used warping wings for circular flight, while Curtiss was forced to develop the aileron method of lateral control, because of prior patent rights controlled by the Wrights. Not only were the methods for control dissimilar, but the control levers were entirely different.

For instance, Curtiss provided directional control by means of a wheel mounted on a column, similar to an automobile wheel. Vertical motion was actuated by a fore and aft movement of the column. A yoke encircled the pilot's shoulders and the machine banked in obedience to the ailerons as the flier leaned to one side or the other, pushing the yoke.

Figure 6. An N-9 trainer returns to NAS Pensacola's seaplane ramp in this vintage postcard. (Author's collection)

Because of the need for actual physical strength to warp the entire wing structure of the Wright machine, more leverage had to be provided. In the two-seater machine, the pilots sat side by side. Between them was installed one vertical lever which controlled the elevator. On the outboard side of each pilot was mounted another vertical lever, making possible dual control. In each case, this lever actuated the wing warping, while at the top of the column was hinged a small secondary lever which, by a twist of the wrist, controlled the rudder.

In civilian aviation circles, an aviator was either a Curtiss flier or a Wright flier, but never both. In addition, a Wright pilot was classified as either a right-handed or left-handed, according to his particular training. Obviously, this condition could not long exist in an organization like the Navy, where interchangeability must be maintained to the highest degree.

It is a matter of record that by 1915 this condition was indeed corrected through an ingenious, but simple, method of stick control developed in France known as the Deperdussin method. In many respects it was the prototype of modern aircraft controls. It consisted of a vertical yoke topped by a wheel; the wheel operated the aileron for banking and circular motion, while the yoke operated the elevators for up and down direction. A rudder bar was added for directional control. After extensive experiments by Lieutenant Pat Bellinger, the Navy settled upon a variation of the basic Deperdussin method very much like the method universally in use today.[16]

However, alarming crash rates caused both U.S. Army and U.S. Navy aviators to grow increasingly dissatisfied with the performance of the Wright aeroplanes. On February 9, 1914 Lieutenant H.B. Post, U.S. Army, was killed when his Wright C seaplane crashed into San Diego Bay during a solo altitude record attempt.[17] Six Army aviators had been killed in this model aeroplane but Orville Wright "...insisted that the accidents were not the fault of the plane but rather the fault of the pilots who had not been properly trained to fly them."[18]

Figure 7. Apparently the N-9 also made an excellent platform for the occasional wing walk. One wonders if this particular individual was the student or the instructor. (Courtesy National Museum of Naval Aviation)

The U.S. Army disagreed, arguing that the aviators involved in the accidents were some of their best pilots. After several more accidents, the Chief Signal Officer directed on February 16, 1914 that the Wright aeroplanes be grounded until further notice.[19]

The U.S. Navy was not far behind. The first Naval aviation fatality had been in a Wright aeroplane and the second in a Burgess aircraft built under a Wright contract. By the time of the arrival of the Naval aviation camp in Pensacola in 1914, Wright aeroplanes had fallen out of favor with Naval aviators, their general consensus that the aeroplanes were untrustworthy.

Further Wright aeroplane designs did not fare any better. The Navy Department purchased the Wright's first true flying boat, the Wright Model G Aeroboat, in 1914 and had it delivered to Pensacola.[20] Testing did not go well and Pensacola recommended that the "Wright two-propeller airboat" be returned to the manufacturers due to serious deficiencies.[21] The Wright brothers' obsessive focus on protecting their patent had left little time for improvements to their aeroplane design. Although they were the first to fly and several years ahead of the world in aeronautical knowledge, they had lost their competitive edge to other inventors with more advanced concepts.

Nonetheless, Pensacola's Naval aviators were ready to stretch their wings. There were a lot of aviation records waiting to be set or broken.

FLYING THE EARLY AEROPLANES

Vice Admiral Bellinger described what it was like to fly the early Wright and Curtiss aeroplanes that were first used in Pensacola:

The Wright hydroaeroplane:

Wright hydro-aeroplane with twin floats...high speed something in the order of 55 miles per hour and in landing it touched the water at a speed of about 35 miles per hour. The pilot and passenger sat on flat seats side by side mounted on the wing; your feet stuck out in front, resting on the outrigger structure of a few pieces of wood. There were no straps to hold you in and the outrigger structure enabled you to clamp your heels against a cross bar of the outrigger structure preventing you from sliding forward when in a steep glide. There wasn't much sense of security for a newcomer and ...the fact that you sat in the clear with nothing in front of you and little around you gave the would be passenger a lot of food for thought before making his final decision to fly.

The Curtiss hydroaeroplane:

...pilot and passenger sat side by side on seats with canvas bottoms, whose sides were a light metal framework extending up almost to the shoulder. To the top of this framework was attached a light leather harness, that went over the shoulder so that the pilot and passenger were partially strapped in, but anyone could get clear by slumping one shoulder and ducking out. This frame work swung to the right and left by leaning the body to one side or the other, and operated a pulley system which in turn operated the ailerons which controlled the banking of the plane.

...The control column worked the horizontal rudders. When shoved forward the horizontal rudders sent the plane down by the nose and when pulled back the plane climbed. The engine throttle was controlled by the toe of the right foot. The foot held the throttle open against spring pressure and when the foot was removed the throttle closed to idling. The ignition switch was easily accessible on the control column.

NOTES

[1] Lieut. Murray, of Aviation Corps, Dashes to Death (1914, February 17). *The Pensacola Journal.* [2] Naval Aviator Falls About 100 Feet, Machine Falling On Him, Crushing the Skull. (1915, May 9). *The Pensacola Journal.*

[3] Student Aviator Plunges 150 Feet To Death. (1916, May 25). *The Pensacola Journal.*

[4] A Military Funeral For Dead Aviator. (1916, May 26). *The Pensacola Journal.*

[5] Bennett, F.M. (personal communications). (1919, March 15). Special Collection, 73-6, Box 2, John C. Pace Library.

[6] Saufley, Breaking His Own Record Falls To His Death On Island. (1916, June 10). *The Pensacola Journal.*

[7] Casey, L.S. (1981). *Curtiss The Hammondsport Era 1907-1915.* New York: Crown Publishers, Inc.

[8] Bellinger, P. (n.d.). *The Gooney Bird.* Unpublished manuscript. Washington, D.C.: Naval Historical Center, Operational Archives Branch. NRS 1982-2.

[9] Bellinger, P. (n.d.). *The Gooney Bird.* Unpublished manuscript. Washington, D.C.: Naval Historical Center, Operational Archives Branch. NRS 1982-2.

[10] Bellinger, P. (n.d.). *The Gooney Bird.* Unpublished manuscript. Washington, D.C.: Naval Historical Center, Operational Archives Branch. NRS 1982-2.

[11] Van Deurs, G. (1966) *Wings for the Fleet.* Annapolis, MD: Naval Institute Press, page 15.

[12] Swanborough, G. & Bowers, P. (1968). *United States Navy Aircraft Since 1911.* London: Putnam Aeronautical Books, page 114.

[13] Van Deurs, G. (1972, November). The N-9. *Air Classics, pages 42-49.*

[14] Van Deurs, G. (1972, November). The N-9. *Air Classics, pages 42-49.*

[15] Aviators Sigh As N-9 Planes Go Up In Smoke. (1927, December 16). *The Pensacola Journal.*

[16] *Introduction to Naval Aviation.* (1946). OPNAV: CNO, page 140.

[17] Hennessy, J.A. (1985). *The United States Army Air Arm.* Washington, DC: Office of Air Force History United States Air Force, page 102.

[18] Hennessy, J.A. (1985). *The United States Army Air Arm.* Washington, DC: Office of Air Force History United States Air Force, page 103.

[19] Hennessy, J.A. (1985). *The United States Army Air Arm.* Washington, DC: Office of Air Force History United States Air Force, page 103.

[20] Wright Flying Boat Bought By the Navy Dept. (1914, April 18). *The Pensacola Journal.*

[21] Two-Propeller Air Boat Didn't Fly Very Good. (1914, July 8). *The Pensacola Journal.*

CHAPTER 8
RECORDS AND EXPERIMENTS IN PENSACOLA

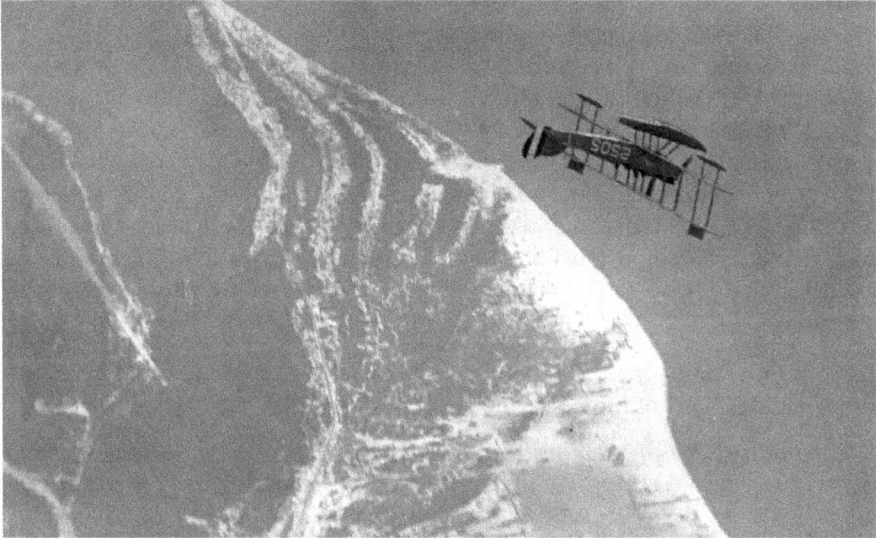

Figure 1. Thought impossible until successfully demonstrated by Captain Francis T. Evans, USMC at Pensacola in 1917, an N-9 seaplane casually loops over the Naval Air Station. (Courtesy Pensacola Historical Society)

EARLY AIRMANSHIP

In a letter dated January 6, 1914, Captain Mark L. Bristol, officer in charge of Naval aviation, provided clear instructions to Lieutenant Commander Henry C. Mustin, Commanding Officer of the Naval Aeronautic Station, on the primary mission of Pensacola's Naval aviators:

The Flying School and the Mississippi form a part of the Naval Aeronautics Station, Pensacola, Fla., as one organization. You will prepare for experiments from the Mississippi but bear in mind, the airmen of the Flying School have several problems yet to solve as airpilots before beginning work as Naval Air Pilots – also work of training airpilots at the school must be pushed first.[1]

The notation "several problems yet to solve" slightly understated the formidable challenges facing Lieutenant John H. Towers, first officer in charge of the Naval Flying School at Pensacola. The most experienced pilot in the U.S. Navy, by 1913 he had amassed more than 151 flight hours and had flown over a quarter of all the flights made in Naval aviation history. Yet even this flight experience was not enough to provide more than a rudimentary understanding of aeroplane performance, and he had to develop an aviation training program from scratch.[2] When asked what they really knew about flying, one of the first Naval aviators replied: "Not much. None of us did. We learned our lessons and made up our rules as we went along."[3]

There were no books to study, no procedures to memorize, and no formal courses of instruction to complete. A young flight student had only the voices of the more seasoned pilots to listen to and learn from. Then the student simply flew to gain practical experience on what to do or not do, and hopefully did not die during the educational process. Pilots flew by the seat of their pants and some thought an aviator who needed an air speed indicator was incompetent. Not that it mattered much, for these early aeroplanes had little or no instrumentation anyway and several years passed before airspeed, attitude, and altitude, the most critical elements of flight, were provided to the pilot. Their non-muffled engines also were incredibly noisy and one aviator noted "If we flew a four-hour hop, we'd be deaf for two hours after shutdown."[4]

Basic aerodynamic principles had not yet been solved and the first aeroplanes reflected this. One of the first Naval aviators noted:

Flying in those days was not too thoroughly understood except in its fundamentals. There were many who held different ideas as to what would happen under certain conditions. The reaction of controls on the plane's flight was another bone of contention.

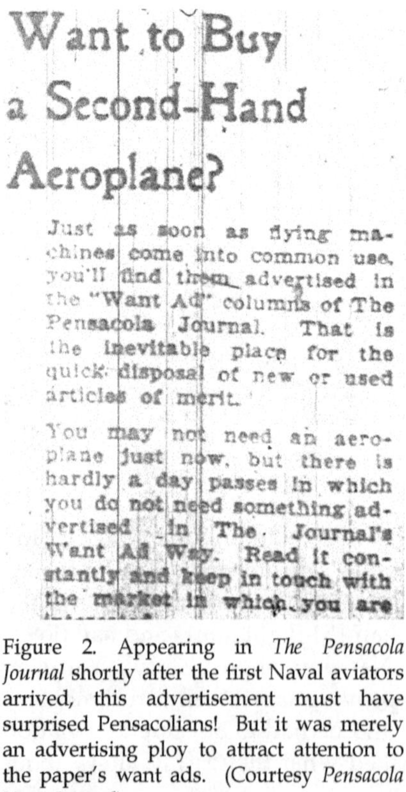

Want to Buy a Second-Hand Aeroplane?

Just as soon as flying machines come into common use, you'll find them advertised in the "Want Ad" columns of The Pensacola Journal. That is the inevitable place for the quick disposal of new or used articles of merit.

You may not need an aeroplane just now, but there is hardly a day passes in which you do not need something advertised in The Journal's Want Ad Way. Read it constantly and keep in touch with the market in which you are

Figure 2. Appearing in *The Pensacola Journal* shortly after the first Naval aviators arrived, this advertisement must have surprised Pensacolians! But it was merely an advertising ploy to attract attention to the paper's want ads. (Courtesy *Pensacola News Journal*)

Actually, the plane had to be flown all the time while in the air, the controls had to make the plane do what it was supposed to do, because the plane didn't naturally do anything correctly. The pilot had to make it. This gradually changed with the developments of aviation. The pilot flew his plane by feel. There were no instruments to aid him. He had to know when he had proper flying speed for his plane to react to the controls. There was nothing to tell him except his feel of the plane. He had to know when the plane was slipping or skidding or stalling.

Climbing to an altitude in those days with those planes was a rather delicate operation, as the control of the plane rested entirely in the judgment of the pilot; there were no instruments, and a wrong estimate by the pilot on the slow side would cause the plane to lose flying speed and probably a fall on one wing and then a spin. [5]

Another early pilot noted:

...fliers spoke of driving, not flying, their bamboo-tailed pushers. These tilting, tipsy "crates" were inherently unstable. Successful driving either on the ground or in the air required constant movement of controls. Few experts and no novices attempted this balancing act in any breeze. Because the air was still near sunset and in the early morning, Curtiss limited his students' practice to those hours.[6]

A hydroaeroplane flier recalled:

The planes of those days particularly the hydro-aeroplanes with the pontoon hanging over the center, were not balanced for flight as training planes of a much later date were balanced. When flying those planes in level flight, pressure was always exerted on the control column to keep the plane from climbing into a stall. Whenever the engine failed, the control column was instantly pushed forward to get the plane headed down in a glide. A rule of thumb was in force that was the Bible for all pilots and students 'Whenever in doubt or trouble, head down immediately.'[7]

While takeoffs with power were mandatory, landing with power available was all too frequently an option. Early engines were unreliable and subject to sudden seizures without warning. Accordingly, pilots were nothing if not well trained to be constantly aware of their surroundings and the immediate availability of an emergency-landing site in the event of a forced landing. For Pensacola's first Naval aviators, who flew only hydroaeroplanes and flying boats, the choice was easy: circle down to a water landing. The options were fewer for those who were caught inland as the Pensacola area was covered with thick forests of pine and oak, offering few hospitable landing areas.

On the bright side, airspeeds were low, allowing most aviators to walk away with minimal injuries from the frequent crashes and forced landings. A crash of one of the U.S. Navy's first aeroplanes was described as the "resounding sound of light woodwork breaking" from which the aviator escaped unharmed.[8]

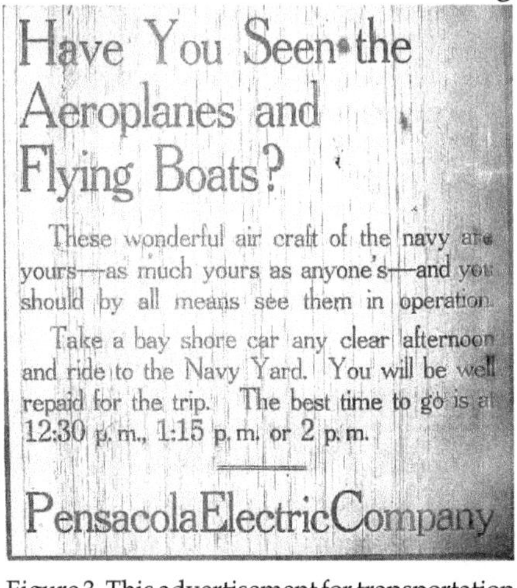
Have You Seen the Aeroplanes and Flying Boats?

These wonderful air craft of the navy are yours—as much yours as anyone's—and you should by all means see them in operation.

Take a bay shore car any clear afternoon and ride to the Navy Yard. You will be well repaid for the trip. The best time to go is at 12:30 p. m., 1:15 p. m. or 2 p. m.

Pensacola Electric Company

Figure 3. This advertisement for transportation to view the aeroplanes at the old Navy Yard indicates that it did not take long for local businessmen to realize potential business opportunities. (Courtesy *Pensacola News Journal*)

"AN AEROPLANE SAILS LIGHTLY OVER THE CITY"[9]

While Pensacolians were not necessarily unaware of the dangerous work facing their new aviator friends, they could certainly appreciate the gracefulness of the aeroplanes that suddenly appeared over their homes and businesses, a view unparalleled in the world. As the citizens of Pensacola's admiration for their Naval aviators increased, so did the pilots in turn take every opportunity to show off their flying skills to their admiring public. *The Pensacola Journal* reported nearly every aeroplane sighting, every incident and every accident with breathless exuberance.

The very sight of one of the U.S. Navy's aeroplanes would cause all business in Pensacola to stop. The following is just one wonderful example of typical early newspaper coverage of the Naval aviators:

Two aeroplanes from the only aeronautical station in the world were piloted gracefully and it appeared leisurely over the city of Pensacola proper yesterday afternoon and were viewed for some time by hundreds of people who were attracted by the whirring of their motors and the long time the two craft remained in the air.

The aviators had the planes under perfect control and hundreds were thrilled at the maneuvers they executed possibly seven or eight hundred feet in the air. The weather was ideal for even this daring sort of work, and citizens who had not thus far had the opportunity of viewing an airship in the city were rewarded to their satisfaction during the afternoon. For a long time the two daring aviators circled over the city and in more than one office work was forgotten for the time being because of the daring displayed by the couple of airmen. The motors made a din when the big machines were gracefully sailed and turned about in the air, the sound giving one the impression of a planing mill in the distance.

At all points of the city were people grouped witnessing the thrilling sight and this was more especially true of the 'home folks' in the upper northern part of the city who perhaps had the clearest view of the maneuvers. Along the bay front the sight of the airmen at work is not so unusual, but the resident portion of the city has thus far not been given the thrilling 'treat' as was handed out yesterday afternoon. And until the maneuvers of yesterday there was no attempt to go over the city proper as was done yesterday afternoon. [10]

Soon after his arrival, Lieutenant Commander Mustin graciously granted permission for his aviators to participate in Pensacola's February 1914 Mardi Gras festivities. *The Pensacola Journal* was quick to note that "this will be one of the best features ever obtained for the entertainment of visitors to the city, thousands of whom have never seen a flying machine of any description and it will be a feature not obtainable by any other city giving a Mardi Gras celebration this year."[11] The aviators and their flying machines were soon in great demand for local celebrations, inaugurations and social occasions.

Figure 4.   Soon after their arrival, Naval aviators gladly participated in Pensacola's Mardi Gras festivities. The caption for this photograph read "King Priscus XVI Arriving In His Royal Aeroplane At Foot of Palafox Wharf to Commence His Reign. " (Courtesy *Pensacola News Journal*)

FIRST PENSACOLIAN TO FLY

Let there be no doubt that Doctor M.E. Quina of Pensacola wanted to fly. During Nels J. Nelson's appearance in Pensacola in 1911, *The Pensacola Journal* noted that "many requests have been made by well-known men to be carried up as passengers. The most persistent of all is Dr. Quina, and it is expected that he will make the attempt next Sunday, if agreeable to the aviator."[12] No record exists to indicate whether Dr. Quina ever flew with Mr. Nelson but on March 3, 1914 Dr. Quina definitely took off in a U.S. Navy aeroplane to become the first Pensacolian to fly.[13] For reasons unknown, Lieutenant Commander Mustin coordinated with Washington to obtain permission for Dr. Quina to make the flight. Front-page news on *The Pensacola Journal*, the good doctor fared well on his first flight, although the paper was quick to note that on landing"...the doctor apparently still had nerve, though it was noticed that he was a little pale."[14]

> MARDI GRAS
>
> **4** BIG PARADES **PENSACOLA** **6** BRASS BANDS
>
> **MARCH 5 to 7**
>
> An excellent opportunity to see Uncle Sam's fighting craft—Cruisers, Submarines, Monitors, Destroyers, and the
>
> ### AEROPLANES
>
> Exhibition flights by Navy Aviators in the finest machines owned by the United States Government will be a daily feature.
>
> Masked Balls, Confetti, Music, Fun and Frolic in the Deep Water City.
>
> YOU MUSN'T MISS IT!
>
> Low rates on all railway and steamer lines
>
> ## COME TO PENSACOLA

Figure 5. Advertisement in *The Pensacola Journal* for one of Pensacola's Mardi Gras celebrations. (Courtesy *Pensacola News Journal*)

The newspaper was also quick to point out that "...it is useless for parties to apply to the captain and aviators for permission to take joy rides, for they will be refused."[15] Despite these dire warnings, some persistent citizens apparently petitioned anyway and were rewarded with 'joy rides.' *The Pensacola Journal* reported on another civilian's flight in 1914:

Max L. Bear, the well known wholesale man, made a flight with Lieut. Chevalier yesterday afternoon, and stated that it was a enjoyable sensation to dash through the air at the rate of 58 or 60 miles an hour. "It is not until you approach the starting point," said Mr. Bear, "that the uninitiated comes to realize there is danger. Up in the air, one does not feel that he is traveling at a high rate of speed, as there are no objects to pass. A man should not try to go up in an airship if he lets anything akin to fear possess him, for it is then one is really in danger. But I will have to confess that air traveling gives one a fine sensation."[16]

MEXICAN CAMPAIGN: FIRST CALL TO ARMS FOR PENSACOLA AVIATORS

Barely unpacked in Pensacola, the first bugle call to action for Naval aviation occurred in April 1914 in support of operations in Mexico.

The Mexican Revolution began in 1910 when Francisco I. Madero called for an uprising against President Porfirio Diaz. Successfully forcing him from

Figure 6. In April 1914 Naval aviators from Pensacola embarked the Curtiss flying boat C-3 (left) and hydroaeroplane A-3 (stored atop the gun turret) aboard the USS MISSISSIPPI for operations in Mexico. (Courtesy U.S. Navy)

office in 1911, Madero ruled until murdered in 1913 when General Victoriano Huerta seized power. Madero's followers rallied behind Venustiano Carranza to oppose Huerta and fighting erupted throughout the country. U.S. President Woodrow Wilson refused to recognize Huerta's government while openly supporting Carranza. Eventually American business interests in Mexico forced the United States to intervene based upon a minor incident. On April 9, 1914 a paymaster from the USS DOLPHIN landed with a boat crew at Tampico for supplies, unaware that martial law had been established and orders issued that no one was to be permitted to land at the port. Forces loyal to General Huerta arrested the men but soon released them with an apology from the local commander. Rear Admiral Henry T. Mayo, Commander of U.S. Naval forces in the region, was not satisfied with this apology and demanded that the United States flag be saluted with special ceremony. General Huerta refused and several thousand U.S. forces supported by an armada of battleships and cruisers were ordered to Mexican waters off Tampico and Vera Cruz. Among these forces were U.S. Marines and Naval aviators from Pensacola.

On April 20, 1914 an aviation detachment of Lieutenant Towers, First Lieutenant Smith and Ensign Chevalier with two Curtiss flying boats and one Curtiss hydroaeroplane embarked aboard the USS BIRMINGHAM (CL 2)[a] for Tampico.

A second aviation detachment of Lieutenant Commander Mustin, Lieutenant (Junior Grade) Bellinger, Lieutenant (Junior Grade) Saufley, Ensign Stolz and Ensign LaMont with the Curtiss flying boat C-3 and Curtiss hydroaeroplane A-3, with a land chassis, were embarked aboard the USS MISSISSIPPI for Vera Cruz.

On April 21, 1914, President Wilson issued orders for American forces to seize the port facilities and customs house at Vera Cruz. The USS MISSISSIPPI arrived at Vera Cruz on April 24, 1914 and the next morning Bellinger and Stolz flew the first scouting missions. On May 6, 1914 during an overland scouting mission, an aeroplane flown by Bellinger with Saufley aboard was hit by rifle fire, the first combat damage to a Naval aeroplane. The USS BIRMINGHAM joined the USS MISSISSIPPI at Vera Cruz and Naval aviators from both detachments flew reconnaissance missions over Mexico with Bellinger at one

---

[a] This is the same USS BIRMINGHAM from which civilian test pilot Eugene Ely made the first takeoff from a ship on November 14, 1910.

point flying for 43 consecutive days. By June 1914, the aviators' services were no longer needed and they returned to Pensacola to a hero's welcome.[b] They had flown several hundred flights without an accident.[17] They also had proven the value of aeroplanes in wartime operations and Secretary of the Navy Josephus Daniels declared: "Aeroplanes are now considered one of the arms of the Fleet the same as battleships, destroyers, submarines, and cruisers."[18]

FIRST NIGHT FLIGHT IN PENSACOLA

The Naval aviators resumed their training and on August 7, 1914 they made the first night flight in Pensacola's history. The affair started unassumingly enough with an 8:30 p.m. takeoff from the waters off the aeronautic station by Chevalier and a mechanic. What happened after they made a quick pass over the city stunned spectators.[19] A gasoline leak caused the flying boat to burst into flames. Chevalier, flying at an altitude of about 800 feet, immediately put the flying boat into a dive, landed, and then he and his mechanic "...leaped from the burning machine as soon as it settled on the water and went some distance away, fearing the gasoline tank might explode."[20] This was only the first of many burning seaplanes, landplanes, and balloons to pass over Pensacola.

FUN IN TOWN

Not that it was all work at the Naval Aeronautic Station. Vice Admiral Bellinger, one of the first Pensacola Naval aviators, felt compelled to mention in his autobiography one adventure that must have received some unwanted publicity:

One Sunday all planes were out heading for an attractive house on the bay shore where we were invited for breakfast. It was a very nice and unique affair given by the Knowles family of Pensacola. I am mentioning it because it has been referred to as an historic occasion, one of the firsts of Pensacola. Five planes on the sandy beach in front of the Knowles house was quite a sight in early 1914.

FIRST NAVAL AVIATION VISIT TO MOBILE

On February 13, 1915, Bellinger and Ensign Bronson flew to Mobile in the Curtiss flying boat AB-3 to participate in Mardi Gras festivities.[21] The USS WORDEN, a torpedo boat destroyer, was stationed 15 miles out along the route to provide assistance if required.[22] As described in *The Mobile Register:*

In the history of the port of Mobile, there was probably no more spectacular an arrival than that of the United States naval hydro-aeroplane K6-25 (sic) at 1:15 Saturday. The flying machine made the flight from Pensacola via Fort Morgan to represent the bureau of naval aeronautics at the Mardi Gras. The hydro-aeroplane followed the channel to a point near the foot of Church street and then circled, dipped and gyrated

---

[b] Carranza's forces forced Huerta to leave the country in July 1914 and Carranza ruled until he was killed in 1920 during a revolt led by General Alvaro Obregon.

LIEUT. BELLINGER AND HIS MACHINE.

Figure 7. Photograph from *The Mobile Register* on February 17, 1915 showing Lieutenant (Junior Grade) Bellinger and Ensign Bronson at Mobile. (Courtesy *The Mobile Register 1915©*. All rights reserved. Reprinted with permission.)

above the business district of Mobile. Although there were thousands of persons scattered along the harbor front to see the arrival of the naval vessels, probably ten thousand additional strained their necks to watch the flying machine circle over the city. After a short flight the machine was guided to the government sheds, just east of the Turner Terminal Docks, where it alighted, and will make its rendezvous while in Mobile. [23]

Bellinger made several additional flights, carrying numerous Mobilians as passengers, until returning to Pensacola on February 19, 1915. For all their excitement, these were not the first flights of an aeroplane over Mobile as some four years earlier Mobile had sponsored the "Great Southern Aviation Meet."

ALTITUDE RECORDS IN PENSACOLA

Despite the occasional setback, Naval aviators now felt confident enough to attack several aviation records. One of the first claimed by a Pensacola-based aviator was an altitude mark for hydroaeroplanes, set by Bellinger in April 1915 when he climbed to 9,500 feet in a Dunne machine. It took him approximately one hour to reach this altitude whereupon he turned off his engine and enjoyed a 15-minute glide to splash down. [24] A climb rate this slow explains precisely why nearly all training flights in Pensacola were made at a height of only a few hundred feet. Nonetheless, the very next day, Bellinger climbed to an altitude of 10,000 feet to further extend his record. [25]

On November 4, 1915 Saufley broke this record when he climbed to 11,000 feet in a Curtiss machine. [26] He bettered this by climbing to 12,050 feet on December 4, 1915; 16,010 feet on March 29, 1916; and 16,072 feet on April 2, 1916. Additionally, on June 5, 1916 Saufley broke the world's record for hydroaeroplane endurance by remaining airborne for eight hours and five minutes. [27]

To the joy of local civic and business leaders, every record received national attention and provided Pensacola unprecedented publicity. As an editorial in *The Pensacola Journal* noted after one of Saufley's records:

Every newspaper in the country told yesterday how Lieutenant Saufley had smashed another aviation record; and every reader talked or thought more or less of Pensacola.

Measured in dollars, this publicity is worth a staggering sum. Measured in pulling power, however, the advertisement is worth a thousand times more.

Other resort places spend fortunes advertising some such attraction as a golf links, a lake, a Lovers Leap rock, or a Sulphur spring. Here in Pensacola we have the greatest advertising feature in the country, and the United States Government is paying for it!

With aeroplane battles raging in the air, and aeronautic appropriations being increased in congress; with preparedness filling the national mind, and the fact becoming more and more certain that the aviation arm of the serivce is our most potent protection, the entire American people are thinking aeroplanes – and to a less degree, Pensacola.

Let us therefore continue to build roads, plant crops, improve our cattle, and boost West Florida. Any news that goes out with a Penacola date line will find eager readers henceforth. Let us send out news that will bring colonists! [28]

Not all record attempts ended happily. On June 9, 1916 while trying to break his own record for sustained hydroaeroplane flight, Saufley stalled and spun to his death onto the sands of Santa Rosa Island. [29]

FIRST FLIGHT OVER MILTON

Bellinger also made the first flight over Milton. [c] He described his flight:

The Secretary of the Navy directed the Commandant to send a plane to Milton, Florida, a little town on a bayou southeast (sic) of Pensacola, if practicable, for the 4th of July, 1915 celebration at that place. I was elected and went in the Burgess-Dunne. I showed off over the town, then landed near the main dock in a small bayou and was taken ashore. I led the procession, sitting beside the driver of a big beautiful red engine. After the parade, I was taken to the ice house to sit in for awhile. Then lunch, after which I manned the plane, made another exhibition flight over Milton, then continued home without incident. [30]

AEROPLANE CONSTRUCTION

Somewhat of a controversy occurred in the fall of 1915 when the Secretary of the Navy announced that experiments in aeroplane manufacturing would take place at Pensacola. [31] As might be expected, civilian aeroplane manufacturers objected strenuously to this perceived encroachment by the government into their business. In a public argument aired in the pages of Aerial Age, President C.E. Dobson of the Pensacola Chamber of Commerce responded to an article criticizing Pensacola's role:

It is not the intention of the Navy Department, as I understand it, to build all its aeroplanes, or even a majority of its aeroplanes, at Pensacola, but it is the idea, I think, to have at Pensacola a plant for experiment and to produce a certain number of aeroplanes, located at the place where the major portion of the naval flying is done, supervised by men who fly themselves, who are seamen and know all the needs. The

---

[c] Milton Airport was not dedicated until June 1935.

needs of the naval aviator in a machine for carrying him above the water are differ-ent from those of a flyer over the land. He wants more horsepower with less weight, for this machine is heavier; he wants capacity for sustained flight, he wants pontoon arrangement which will allow him to alight on or arise from the surface of the water; he wants a general structure that will best suit storing aboard and launching from a naval vessel.

Will he get from the private manufacturer the study, investigation and patient experi-ment necessary to meet these needs? Past history does not indicate it. Why, then, should not the Government experiment for itself and construct for itself; not entirely, but as a leader, a pacemaker, for private manufacturers? Judging by the past and pres-ent, private manufacturers need such a pacemaker. There is not a motor in this coun-try adapted for use in seaplanes that can be relied upon to run for five hours without breaking down. They should run double or triple that time.

There is not a motor made in this country that, under naval tests, will develop over eight-five per cent of the horsepower claimed for it. There has been too much desire to reap the harvest and too little to do exact and painstaking work. There have been machines furnished the Navy which could not be wet up until they were partially built over. There have been cracked castings, badly balanced machines, parts that would not fit, shafting that broke under the first stress. In fact, it has been made as an official recommendation time and again by one of the leading naval aviators that no more machines be purchased from a certain prominent manufacturer until he took steps to remedy the almost criminal carelessness with which his machines were constructed.[32]

FIRST NAVAL AEROPLANE BUILT IN PENSACOLA

Figure 8. This aeroplane is identified as the first aeroplane to be completely built at NAS Pensacola. It was designated the McDonnell 152 in honor of its designer, Lieutenant E.O. McDonnell. (Courtesy Pensacola Historical Society)

While nothing further is documented of this controversy, a year later a newspaper article did report on an aeroplane being built at the station:

The first airplane to be built at the aeronautic station here is now under construction and is expected to be completed within six weeks. It was designed by Lieut. E.O. McDonald (sic) and the plans and specific jobs have been approved by the navy department.

Some delay was experienced when the airplane was started because of the lack of skilled labor, and the first work has just commenced.[33]

Reportedly, McDonnell[d] lost enthusiasm for his own project and left others to finish it after his transfer. No further record was found indicating whether or not the aeroplane flew successfully.

EXPERIMENTS

In addition to student training, Naval aviators at Pensacola also were conducting various aviation experiments. The risk was sometimes great. As one officer recorded, "Various types of machines were tried out in the process of developing an efficient and safe type, and some conclusions were arrived at only as the result of crashes in which officers were killed who had done much to promote interest in Naval aviation."[34]

A report by the Secretary of the Navy in late 1915 provides a good overview of the types of experiments carried out in Pensacola:

Experiments of all kinds are continually being carried out at the aeronautic station. Some of these experiments can be combined with the flying school work, such as efficiency of different lubricating oils and of different grades of gasoline; the preliminary test of different types of floats for aeroplanes; tests of instruments; tests of special life preservers for aviators; experiments with special clothing and various other tests. In addition to the qualified aviators who have been appointed, navy air pilots have tested out new types of floats for aeroplanes in rough weather in the Gulf of Mexico; tested a launching device that has been installed on the North Carolina, and used to launch an aeroplane from that ship under way; have made long flights to test the reliability of motors and the endurance of aviators; tested aeroplanes for speed and climb; are testing bombs and bomb-dropping appliances; trying out safety straps in all kinds of weather to demonstrate their stability and reliability; conducting running motor and propeller tests, operating special hoisting apparatus required for aeroplanes on the North Carolina, and a large number of minor experiments.[35]

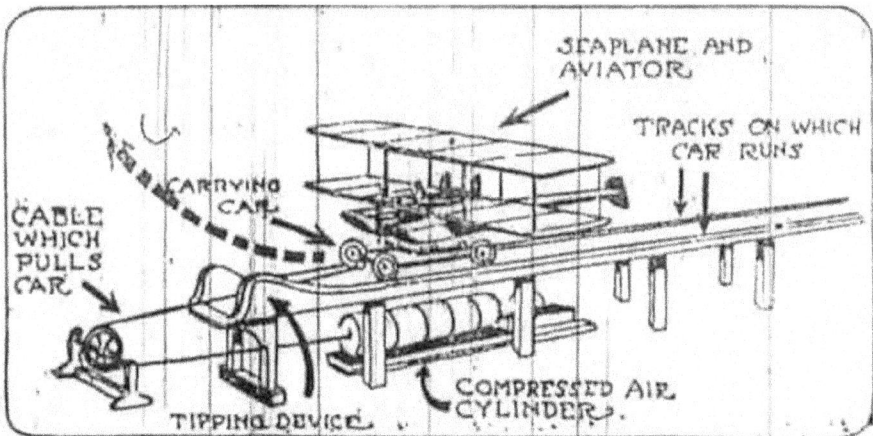

Figure 9. A May 20, 1917 graphic in *The Pensacola Journal* explained how a heavier-than-air flying machine could be hurled from a ship at a speed sufficient to allow the machine to fly without touching the water. Naval aviators in Pensacola made the first successful catapult launches from ships at sea. (Courtesy *Pensacola News Journal*)

---

[d] Lieutenant McDonnell was holder of the Congressional Medal of Honor and retired as a Vice Admiral.

CATAPULT TESTING

Some of the most important experiments in Pensacola investigated methods of integrating the aeroplane with the fleet. Devising a method of successfully launching an aeroplane from a ship was a top priority. Three options presented themselves. The first, hoisting an aeroplane off and on a ship, had been successfully demonstrated in 1911 but this evolution required the ship to stop, generally considered unacceptable, especially during combat operations. The second option, landing and relaunching an aeroplane from a ship, also had been successfully demonstrated in 1911 but this required the building of a platform aboard ship which masked the ship's guns. The third option, which never had been done, was to catapult an aeroplane off a moving ship and then recover the flying machine when time permitted.

Experiments with shipboard catapults had been conducted as early as July 31, 1912 when Lieutenant Theodore Ellyson, Naval Aviator Number 1, unsuccessfully attempted to launch by catapult in the A-1, Naval aviation's first aeroplane, at Annapolis, Maryland. On November 12, 1912 Ellyson did make the first successful launching of a hydroaeroplane by a catapult at the Washington Navy Yard. In December 1912 a flying boat was successfully launched from the same catapult. All these launches were made from shore. Further experiments to launch aeroplanes from ships would occur in Pensacola several years later.

On April 16, 1915 Lieutenant Bellinger made the first successful catapult launch at sea in the AB-2 flying boat when he was shot from a stationary coal barge in Pensacola Bay.[36] Naval aviators were now ready to mount a catapult aboard a ship.

Figure 10. On April 16, 1915, Lieutenant Bellinger was shot from Coal Barge No. 214 stationary in Pensacola Bay. (Courtesy National Museum of Naval Aviation)

On November 5, 1915 Lieutenant Commander Mustin made the first catapult launch from a ship when he was shot from the stern of the USS NORTH CAROLINA (ACR-12)[e] in Pensacola Bay.[37] Mustin later decribed the sensation of a catapult shot as "...being something like the sensation one gets in the pit of the stomach when an elevator drops, only more so."[38] Lieutenant Bellinger also made a catapult shot after he had "...been to a party the night before, and I can testify that the jolt of a catapult launching clears one's head."[39]

---

[e] The USS NORTH CAROLINA had replaced the USS MISSISSIPPI as aviation station ship when the latter was transferred to Greece in July 1914.

Figure 11. On November 5, 1915 Lieutenant Commander Mustin was shot from the USS NORTH CAROLINA's stern in Pensacola Bay. (Courtesy National Museum of Naval Aviation)

Additional successful catapult launches were made in the succeeding days. On November 8, 1915, First Lieutenant Cunningham attempted the first catapult launch by a Marine aviator.[40] Unfortunately, he crashed during take-off and further tests were delayed for several months.

Finally, on July 12, 1916, the last milestone in the catapult tests was passed when Lieutenant Chevalier was shot from the USS NORTH CAROLINA while the ship was underway in Pensacola Bay.

Experiments continued, but upon the entrance of the United States into World War I, all catapults and aeroplanes were removed from the USS NORTH CAROLINA

Figure 12. On July 12, 1916, Lieutenant Chevalier was shot from the USS NORTH CAROLINA underway in Pensacola Bay. (Courtesy National Museum of Naval Aviation)

and other ships to prevent interference with convoy escort duties.[41] It was not until after the war ended that an interest in embarking shipboard aeroplanes was revived.

FIRST LOOP IN A SEAPLANE AND FIRST SPIN RECOVERY

One of the more remarkable aviation achievements recorded in Pensacola was one accomplished by U.S. Marine Corps aviator Captain Francis T. Evans.

Pilots recognized very early that "speed was life" but did not know what caused a stall, or the correct recovery procedure from an unintentional stall and resultant spin, the number one cause of aircraft mishaps even to this day. An early Naval aviator acknowledged:

...few people understood the 'deadly nose dive...' Later, men learned that a stalled pusher seldom spun. It usually pushed its nose nearly straight down. The pilot suddenly saw the landscape coming straight at his face. If instinctively he pulled the controls to level off, he was finished, for then the hard-up flippers kept the stalled machine from gaining the necessary speed to start flying again.[42]

If the deadly trait of the old bamboo tailed pusher aeroplanes was to fall over into an uncontrollable nosedive, the new N-9 tractor's equally disagreeable habit when stalled was to whip into a spin, another early Naval aviator recalled:

No one knew how to get out of a spin. For some months, all spins ended in crashes, many of them fatal. Aviation periodicals forgot their "deadly nose dive" and spoke often of "the deadly tail spin." In the winter of 1916-1917, some Pensacola instructors argued that since it had a fuselage and stabilizers, an N-9 might do stunts impossible in the old pushers. Some thought it might even loop. Others believed the heavy pontoon would tear it apart if a loop were attempted.[43]

On February 13, 1917 Evans answered both questions. On this day he not only performed the first loop of an N-9 seaplane, but he also forced it into a spin and then successfully recovered.[44] Of the two remarkable achievements, the more important one was not that he had successfully looped a seaplane, something thought not possible[f], but that he had recovered from a spin, which no American had ever done before, even by 1917.[g]

Evans discovered the solution accidentally when he stalled and spun during his first loop attempt. Without realizing that he was in a spin, he pushed the nose over to gain airspeed, applied opposite rudder and the aeroplane recovered. Once he realized what he had done he tried several more successful recoveries. Certain that his fellow aviators would not believe him without a demonstration, he returned to the Naval Aeronautic Station and from an altitude of 2,000 feat repeated his feat above the station's flying beach, much to the astonishment of his fellow instructors. His spin recovery technique was immediately incorporated into the flight-training program and he was sent on a demonstration tour to other training sites. He was later awarded the Distinguished Flying Cross "for this contribution to the science of aviation."[45]

Figure 13. A March 9, 1935 issue of *ARGOSY* featured a fictional novelette set at Naval Air Station Pensacola whose opening sentence read "Red Morel, the Navy's only veteran ace, successfully led all others until he directed those catapult experiments – then there was mechanical death to thwart." (Author's collection)

---

[f] Instructors at Pensacola did not believe that the N-9's large pontoons would let it loop or, if attempted, whether the airspeed required for entry would pull the wings off.

[g] An Englishman by the name of F.A. Lindemann is credited with solving the secret of stall/spin recovery in 1914 but the information was not shared for several years out of "military necessity."

A month later an article in *The Pensacola Journal* mentioned that legislation had been introduced in Massachusetts by the President of the Aero Club of New England prohibiting "looping the loop" in that state. He felt that it was "Foolish to risk the lives of such necessary men in operations from which nothing is gained."[46]

While the article mentioned that "It is understood that the navy officials generally discourage attempts at trick flying," Captain Evans' success served to inspire other Naval aviators to attempt the feat. The day after Captain Evans' successful flight, Navy Lieutenant E.O. McDonnell, not willing to be out-flown by a Marine, went out and looped an N-9 twice. Even Bellinger revealed in his autobiography that once he found out about the successful loop he wanted to try one immediately. So he:

...took an N-9, buckled myself in securely in the seat, climbed to three thousand feet, and then headed into a glide with full throttle. When my speed seemed to be enough, I hauled her up into a loop. The plane acted nicely, made the loop, and came out of the loop easily. By this time, I felt very confident, in fact, unduly confident. I thought I should make more loops and try other stunts. But the motor developed a miss, so I called it a day and came on down. [47]

MEDALS OF HONOR AWARDED AT PENSACOLA

Few people realize that at least two Medals of Honor were awarded for actions that occurred in Pensacola. On September 25, 1918 Chief Machinist Mate Francis E. Ormsbee, U.S. Navy was flying an N-9 seaplane when he observed a sister ship crash into Pensacola Bay.[48] He immediately landed near the crashed aeroplane and dove into the water to rescue the pilot and observer. He saved the passenger but was unsuccessful in saving the pilot despite repeated dives into the underwater wreckage. Awarded the Medal of Honor for his heroism, he later qualified as Naval Aviation Pilot Number 25 and went on to a successful career with Pan-American Airways and the Department of Commerce.

The second Medal of Honor for aviation actions in Pensacola was awarded to a U.S. Marine. On February 11, 1921 Private Albert J. Smith was standing sentry duty at Naval Air Station Pensacola when a seaplane crashed near his post.[49] Ignoring the fire from an exploding fuel tank, he pulled the pilot, who was pinned beneath the burning wreckage, to safety. He sustained burns on the head, neck and both hands and was awarded the Medal of Honor for his heroism.

Figure 14. Several years after receiving the Medal of Honor for actions at Pensacola, Pan-American Airways pilot Captain Frank Ormsbee helped save several members of a foundering ship off the coast of Columbia, as documented in this front page illustration. (Courtesy *Pensacola News Journal*)

FIRST CROSS-COUNTRIES

As confidence in engine reliability and piloting skills improved, Naval aviators stretched the geographic limit of their tether. In a "said to be unprecedented" flight, on October 28, 1918 three hydroaeroplanes flew from Philadelphia to Pensacola.[50] A report on their itinerary gives a good idea of the physical demands placed on pilots of that era:

Left Philadelphia at 9:30 o'clock Thursday morning reached Moorhead, N.C. at 12:17 and left Moorhead at 2:12 and reached Paris Island, S.C. at 2:40; reached St. Augustine between three and four o'clock the same afternoon remained in St. Augustine Thursday night and Friday, leaving Saturday, morning between 10 and 11 o'clock, reaching Miami at 12:15 o'clock; reached Tampa at 2:45 Saturday afternoon, spending Saturday and Sunday night there and reaching Pensacola at 2:33 Monday afternoon.

The trip was made along the coast line and the men were thoroughly fatigued on their arrival here, and Mr. Hutchins was temporarily deafened from the roar of the great motors.

While no official information regarding the purpose of the trip was available, it is understood to have been experimental, to determine the feasibility of bringing new machines from the East to Pensacola under their own power. The success of the flight is expected to result in many machines being brought here in this way.[51]

A newspaper editorial a few weeks later commented that this flight "Demonstrated conclusively the practicality of the aerial mail route, and it is a safe prediction that Pensacola will be one of their first cities to receive daily mails through air-ship service."[52]

On February 3, 1919 two U.S. Army Air Service pilots made the "first aerial voyage between Tallahassee and Pensacola" flying Curtiss aeroplanes.[53] As noted in *The Pensacola Journal:*

The planes came originally from Arcadia and are enroute to New Orleans mapping out an air route between the southern army flying school and the Mississippi mouth port. They made a fast flight between DeFuniak Springs and Pensacola, requiring but one hour and twenty minutes for the trip.

The machines are expected to continue their voyage to New Orleans via Mobile, as soon as the weather clears enough for a safe flight. The fog and low clouds yesterday necessitated the planes making the trip between the capital and Pensacola at a height of about 250 feet most of the way.[54]

Other record breaking cross-country flights also followed but within a few years they became commonplace and to the disappointment of spectators, uneventful. In 1925 after two U.S. Marine Corps aviators flew two DH-4Bs from San Diego to Pensacola in a six-stop flight that drew little attention, The Pensacola Journal noted that "Trans-continental flying has reached such a stage that it no longer attracts the publicity and attention it formerly did."[55]

VICTORY LOAN DRIVES

Several visits to Pensacola by U.S. Army Air Service pilots occurred in support of World War I Liberty Loan and Victory Loan Drives. What is unique about these aeroplane stunts is that they would be flown in landplanes and they would set down at either the Naval Air Station or Fort Barrancas, neither of which had a formally established landing field.

The first visit occurred on May 2, 1918 when U.S. Army Air Service Lieutenant Harry W. Thompson, a native of Milton, arrived in his Curtiss JN-4D. Departing from Park Field at Memphis, Tennessee, he initially landed in the "...great two-hundred acre oat field of the Pace Company" before stunting over Milton and Bagdad to demonstrate his flying skills to friends and family.[56]   He then landed on the parade grounds of Fort Barrancas and spent the night preparing for his flight over Pensacola the next day. It was spectacular:

Figure 15. This drawing in *The Pensacola Journal* on December 26, 1916 noted that "The airplane hat is so new to our unaccustomed eye it seems odd -- dare we say freaky. But it will be the semidress hat of late winter." (Courtesy *Pensacola News Journal*)

Pensacolians long used to airplanes stood agape and breathless for twenty minutes yesterday afternoon while Lieutenant Harry Thompson, of the aviation section U.S. Army, looped, nose-dived, tail-spun, and spiraled in the fleck blue of the sky, while thousands of spectators craned their necks below to witness the performance.

At about 3:30 a squadron of airplanes appeared off the city, but failed to attract any attention, as it is an hourly occurrence. Then a speedy canary-colored machine slipped away from the others and poised for an instant over San Carlos court as though to make its introduction bow to the audience before the act.

It hung, almost motionless clearly outlined against the blue vault, the white, red, and blue of its insignia easily visible at that altitude, and then seemed to crumple and fall, like a bird winged in mid-flight. It turned over and over lazily but inevitably falling. 'He's wrecked' the cry went up, and some even sought shelter to prevent injury by the falling plane.

And then under the deft hands of its pilot the machine righted itself and climbed to its former pinnacle. The driver, if he could see the scurrying crowd, must have smiled behind his helmet.[57]

Figure 16. A visiting U.S. Army Victory Loan aeroplane and U.S. Navy airship share the balloon field at NAS Pensacola. (Courtesy Pensacola Historical Society)

A second visit occurred in April 1919 when U.S. Army Air Service fliers from Montgomery, Alabama flew to the Naval Air Station in a visit coordinated by now former Army aviator Harry W. Thompson.[58] Their flight was described:

Immerman (sic) turns, barrel rolls, loop the loop, tail spins and slide slips thrilled Pensacola yesterday, when two great army planes from Montgomery gave an exhibition of nearly an hour, directly over the city, with Lt. J.C. Nall, Lt. A.T. Steele, Lt Harry Thompson and Expert Mechanician J.C. Tomlin taking part in the flying circus.[59]

FIRST WOMEN TO PILOT AN AEROPLANE OVER PENSACOLA

There are two claims for the honor of the first woman to pilot an aeroplane over Pensacola.

As reported in *The Pensacola Journal* in September 1930, the first woman from Pensacola to pilot an aeroplane was Eloise Turner. She was studying flying at the Spartan School of Aeronautics in Tulsa, Oklahoma. The school planned on a flight to Pensacola but there is no documentation that they ever arrived although she may have returned at another time.

As reported in *The Pensacola Journal* in April 1933, Mrs. Ernestine Green made a solo flight from Old Corry Field to become the only woman known to have made her solo hop at Pensacola.

FIRST FLIGHT OF A WOMAN OVER PENSACOLA

On April 28, 1919 during these same Victory Loan flights, Thelma DeBroux "...a charming young yoemanette (sic), who had been most active in the Victory drive at the station" was invited to fly in one the U.S. Army's flying machines.[60] *The Pensacola Journal* noted that "...Miss DeBroux has the distinction of being the first young woman who has made a flight in an air plane, and found the experience pleasant as well as thrilling."[61] DeBroux's title of yeomanette indicates that she was on active duty with the U.S. Navy.

Other women also flew as passengers in U.S. Navy aeroplanes that would not stand the slightest chance of approval today. Imagine this scenario as reported in a 1923 gossip column in the *Pensacola News Journal*:

You should have seen all the Navy wives at the Air Station Thursday morning, waiting for their hops. The instructors were testing the Martins preliminary training plane, and took their wives up with them. Two ladies, I won't mention any names, wore sickly grins, after having looped, split S's, whatever that is, and goodness knows what else. Mrs. Rary and her guest, Miss Callahan also had a hop, just plain flying, no thrills.[62]

CURTISS MARINE TROPHY

As air race fever swept and enthralled the nation from the late 1910s through the 1930s, the U.S. Navy encouraged its pilots to compete in such events as the

Schneider Cup seaplane races, Thompson Trophy landplane races, Pulitzer Trophy international speed races and the Bendix Trophy cross-country races. Rear Admiral William A. Moffett, Chief of the Bureau of Aeronautics, publicly justified the U.S. Navy's participation in these races as a means of testing high-performance aeroplanes and engines to improve the design of combat aeroplanes. Privately, however, his intention was to spark public interest in Naval aviation and to challenge an increasingly presumptuous Army aviation program.[63] As a result, Naval aviators and Army aviators went head-to-head in fiercely contested, but mutually respected, competitions. Naval aviators from Pensacola participated in many races either as contestants or observers, often flying their aeroplanes from Pensacola to the competition site.[64] In one case, however, a Naval aviator won a competition while flying at Pensacola.

On December 30, 1918 Lieutenant T. C. Rodman, USNRF, won the Curtiss Marine Trophy for "...covering the greatest mileage in 10 consecutive hours."[65] Carrying 11 passengers (credits for extra miles were allowed for passengers) in an H-16, he was credited for 842 nautical miles, with only one stop for fueling.

TEST PILOTS

Naval aviation aerial achievements unfolded at a rapid pace as Pensacola aviators and Naval engineers continued to experiment with improved aeroplanes, engines, structural materials, and aeroplane instruments. For years, Pensacola tested nearly every aeroplane considered for Naval service and it was well known that "...acceptance or rejection of new types depend upon recommendations of the officers of the local station."[66] Accordingly, aeroplanes, often accompanied by aeroplane manufacturers and their own test pilots, arrived in a steady flow at Pensacola and it was not unusual to have several manufacturers on station at the same time.

One hydroaeroplane of note flown in Pensacola was the tailless, swept back Burgess-Dunne, seemingly ahead of its time. A pusher, two aviators were seated in tandem, with the wings swept back to such an extreme angle that it looked like a "... great 'v' flying through the air."[67] English Army officer Lieutenant John William Dunne designed and built this aeroplane "dedicated to the principle of inherent stability."[68]

Figure 17. The swept-wing Burgess-Dunne hydroaeroplane ashore at NAS Pensacola. (Courtesy National Museum of Naval Aviation)

W. Starling Burgess of the Burgess Company purchased the American construction rights from Dunne and built a hydroaeroplane version. Test flights were glowing and reflected a unanimous opinion that "...the Burgess-Dunne solves the problem of inherent stability and represents an important development in American aeronautics." The first U.S. Navy model, designated the AH-7, was delivered to Pensacola in early 1916.

However, a problem soon developed that the inherent stability so prized by Dunne was incompatible with military operations. As noted by one expert:

Maneuverability, ease of handling and superior performance determined an airplane's acceptability by the military. Dunne's craft were relatively inefficient compared to conventional aircraft of equal horsepower; excessive stability did not result in ease of control and maneuverability. A reasonable compromise between control and stability was required.[69]

Due to the distance of Pensacola from the principal aeroplane factories, on January 1, 1918 a decision was made to move the department tasked with evaluating experimental aeroplanes to Naval Air Station, Hampton Roads, Virginia.[70]

As we conclude this chapter, we find hydroaeroplanes and flying boats slowly making credible headway into becoming the mainstay of Naval aviation's striking arm at sea. Participation in fleet maneuvers and exercises demonstrated the bombing, sea scouting, photography, and wireless transmission capabilities of aeroplanes. Further, a who's who of Naval aviators continued to pass through Pensacola for their initial flight training before moving on to command squadrons, carriers, task groups and fleets.

While we wait for landplanes to make their first appearance in the U.S. Navy, another form of aviation was taking to the sky over Pensacola.

Figure 18. Given the deafening noise of early aeroplanes, this 1919 experiment with an early aeroplane detection system at Pensacola may have been quite successful. (Courtesy T.T. Wentworth, Jr. Collection, Western Florida Historic Preservation, Inc.)

NOTES

[1] Air Activities Here Launched Early in 1911. (1936, October 11). *The Pensacola Journal.*

[2] Reynolds, C.G. (1991). *Admiral John H. Towers.* Annapolis, MD: Naval Institute, page 69.

[3] Bellinger, P.N.L. (1961, August). Sailors in the Sky. *National Geographic,* 120, pages 276-296.

[4] Rausa, R. (1976, August). Turntable and Traps. *Naval Aviation News,* pages 9-19.

[5] Bellinger, P. (n.d.). *The Gooney Bird.* Unpublished Manuscript. Washington, D.C.: Naval Historical Center, Operational Archives Branch. NRS 1982-2.

[6] Van Deurs, G. (1978). *Anchors in the Sky.* San Rafael, CA: Presidio Press, page 72.

[7] Bellinger, P. (n.d.). *The Gooney Bird.* Unpublished Manuscript. Washington, D.C.: Naval Historical Center, Operational Archives Branch. NRS 1982-2.

[8] Bellinger, P. (n.d.). *The Gooney Bird.* Unpublished Manuscript. Washington, D.C.: Naval Historical Center, Operational Archives Branch. NRS 1982-2, page 15.

[9] An Aeroplane Sails Lightly Over the City. (1914, March 4). *The Pensacola Journal.*

[10] Air-Craft Are Piloted Over City Long Time. (1914, April 1). The Pensacola Journal.

[11] Flights By Aeroplanes Will Feature Mardi Gras Carnival. (1914, February 16). *The Pensacola Journal.*

[12] Fine Flights Witnessed by Many People. (1911, December 1). *The Pensacola Journal.*

[13] Dr. Quina First Pensacolian to Make Flight in Aeroplane. (1914, March 4). *The Pensacola Journal.*

[14] Dr. Quina First Pensacolian to Make Flight in Aeroplane. (1914, March 4). *The Pensacola Journal.*

[15] Dr. Quina First Pensacolian to Make Flight in Aeroplane. (1914, March 4). *The Pensacola Journal.*

[16] Local Aviator Made A Record In High Flying. (1914, August 2). *The Pensacola Journal.*

[17] Four Hydro-Aeroplanes to Swell Navy's Flying Squadron At Pensacola. (1914, July 5). *The Pensacola Journal.*

[18] Reynolds, C.G. (1991) *Admiral John H. Towers.* Annapolis, MD: United States Naval Institute, page 80.

[19] Navy Flying Boat Catches On Fire In Air; Is Destroyed. (1914, August 8). *The Pensacola Journal.*

[20] Navy Flying Boat Catches On Fire In Air; Is Destroyed. (1914, August 8). *The Pensacola Journal.*

[21] Thousand View Seaplane As She Flies Over City. (1915, February 14). *The Mobile Register.*

[22] Makes Flight Over To Mobile. (1915, February 14). *The Pensacola Journal.*

[23] Thousand View Seaplane As She Flies Over City. (1915, February 14). *The Mobile Register.*

[24] Dunne Machine Goes to Height of 9,500 Feet. (1915, April 23). *The Pensacola Journal.*

[25] New Altitude Record Made by Local Aviator. (1915, April 24). *The Pensacola Journal.*

[26] New Altitude Record Made at Pensacola. (1915, November 4). *The Pensacola Journal.*

[27] Saufley Breaks Another World Aeroplane Record, Remaining in Air Eight Hours, Five Minutes. (1916, June 6). *The Pensacola Journal.*

[28] Pensacola's Opportunity. (1916, June 7). *The Pensacola Journal.*

[29] Saufley, Breaking His Own Record Falls to His Death On Island. (1916, June 10). *The Pensacola Journal.*

[30] Bellinger, P. (n.d.). *The Gooney Bird.* Unpublished manuscript. Washington, D.C.: Naval Historical Center, Operational Archives Branch. NRS 1982-2, page 109.

[31] Defends Navy Policy of Developing Yards. (1915, October 20). *The Pensacola Journal.*

[32] Defends Navy Policy of Developing Yards. (1915, October 20). *The Pensacola Journal.*

[33] Airplane Is Being Built At Station. (1916, December 16). *The Pensacola Journal.*

[34] Bennett, F.M. (personal communications). (1919, March 15). *Historical Information, Collection of.* Special Collections, 73-6, Box 2, John C. Pace Library.

[35] Many Flying Machines Are Being Built. (1915, December 13). *The Pensacola Journal.*

[36] Coletta, P.E. (1985) *United States Navy and Marine Corps Bases, Domestic.* Westport, CN: Greenwood Press.

[37] An Aeroplane Is Shot From Ship's Deck. (1915, November 6). *The Pensacola Journal.*

[38] An Aeroplane Is Shot From Ship's Deck. (1915, November 6). *The Pensacola Journal.*

[39] Bellinger, P.N.L. (1961, August). Sailors in the Sky. *National Geographic,* pages 276-296.

[40] Johnson, E.D. (1977). *Marine Corps Aviation: The Early Years.* Washington: Headquarters, U.S. Marine Corps.

[41] Larkins, W.T. (1996). *Battleship and Cruiser Aircraft of the United States Navy.* Atlen, PA: Schiffer Publishing, page 10.

[42] Van Deurs, G. (1966) *Wings for the Fleet.* Annapolis, MD: Naval Institute Press, page 148.

[43] Van Deurs, G. (1972, November). The N-9. *Air Classics,* pages 42-49.

[44] Records Broken When Capt. Evans Loops The Loop In An Hydroaeroplane. (1917, February 14). *The Pensacola Journal.*

[45] Grossnick, R.A. (Ed.) (1995). *United States Naval Aviation, 1910-1995.* Washington, D.C.: U.S. Government Printing Office.

[46] Loop the Loop in Airplane? No! (1917, March 16). *The Pensacola Journal.*

[47] Bellinger, P. (n.d.). *The Gooney Bird.* Unpublished Manuscript. Washington, D.C.: Naval Historical Center, Operational Archives Branch. NRS 1982-2.

[48] Kelly, H.H. & Riley, W.A. (Eds.) (1995). *Enlisted Naval Aviation Pilots.* Paducash, KY: Turner Publishing Company.

[49] Interim Awards of Medal of Honor 1920-1940. Retrieved from http://www.army.mil/cmh-pg/mohint5.htm.

[50] Aviators End Long Coast Line Flight. (1918, October 29). *The Pensacola Journal.*

[51] Aviators End Long Coast Line Flight. (1918, October 29). *The Pensacola Journal.*

[52] Aerial Routes. (1918, November 25). *The Pensacola Journal.*

[53] First Aerial Voyage Made To Pensacola. (1919, February 4). *The Pensacola Journal.*

[54] First Aerial Voyage Made To Pensacola. (1919, February 4). *The Pensacola Journal.*

[55] Marine Fliers End Remarkabl (sic) Distance Jaunt. (1925, August 25). *The Pensacola Journal.*

[56] Milton Boy To Fly, Boosting Liberty Loan. (1918, May 2). *The Pensacola Journal.*

[57] Hundreds See Army Plane In Daring Flying. (1918, May 3). *The Pensacola Journal.*

[58] Flying Will Feature Loan Drive Today. (1919, April 28). *The Pensacola Journal.*

[59] Flying Circus Gives Thrills To Big Crowds. (1919, April 29). *The Pensacola Journal.*

[60] Flying Circus Gives Thrills To Big Crowds. (1919, April 29). *The Pensacola Journal.*

[61] Flying Circus Gives Thrills To Big Crowds. (1919, April 29). *The Pensacola Journal.*

[62] Eleventh Hour Trip To Altar, Navy Wives' Flights; 1862 Love Story. (1923, November 18). *The Pensacola Journal.*

[63] Trimble, W.F. (1994). *Admiral William A. Moffett.* Washington, DC: Smithsonian Institution Press, page 120.

[64] Two Pensacola Planes Start Tuesday For St. Louis Races. (1923, September 23). *The Pensacola Journal.*

[65] An Historical Note. (1930, November 20). *Air Station News,* Pensacola, Florida.

[66] Martin Plane Taking Tests At Navy Yard. (1923, November 18). *The Pensacola Journal.*

[67] Gould, B. (1966). The Burgess Story. *American Aviation Historical Society Journal, 11(4),* page 272.

[68] Wooldridge, E.T. (1983). *Winged Wonders.* Washington, DC: Smithsonian Institution Press, page 16.

[69] Wooldridge, E.T. (1983). *Winged Wonders.* Washington, DC: Smithsonian Institution Press, page 16.

[70] Bennett, F.M. (1919, March 15). *Historical Information, Collection of.* Special Collections, 73-6, Box 2, John C. Pace Library.

CHAPTER 9
BALLOONS AND KITES

Figure 1. Detail from a Pensacola postcard showing "One of the U.S. Navy Observation Balloons In Flight." (Author's collection)

HISTORY OF LIGHTER-THAN-AIR

Lighter-than-air aviation was born on November 21, 1783 when the first flight in history of a manned balloon took place in Paris, France in a Joseph and Etienne Montgolfier designed "aerostatic machine."[1] Powered by a fire built under an alum coated linen envelope, it preceded the Wright brothers first flight of a heavier-than-air machine by almost exactly 120 years.

Only a few weeks later, on December 1, 1783, their remarkable flight was surpassed when fellow Parisian Professor J. A. C. Charlesa made the first manned flight of a hydrogen-filled balloon in a self-entitled "charliere." Charlesa had combined English chemist Henry Cavendish's discovery that hydrogen gas was lighter than atmospheric air with the French brothers A-N and M-N Robert's discovery of a method of coating silk with rubber to form a gas-proof envelope. [2]

The first balloon ascent in the United States was made in a captive balloon constructed by Peter Carnes in Baltimore, Maryland. He weighed too much to fly in his own balloon so a much lighter teenager named Edward Warren flew in his place at Baltimore, Maryland on June 4, 1784.[3]

The first free balloon voyage in the United States was made by Jean-Pierre Francois Blanchard on January 9, 1793.[4] With President George Washington watching, he launched from the grounds of the Walnut Street Prison in his hydrogen-filled balloon, drifted southeasterly, and landed near the town of Woodbury, New Jersey, 46 minutes and 15 miles later.

Since the courses of free balloons were determined only by the "whims and vagaries" of wind currents, lighter-than-air's next goal was controllable flight. It is another Frenchman, Henry Giffard, who is credited with creating the first practical dirigible, or steerable balloon. On September 24, 1852 powered by a steam engine driving a three-blade propeller, he made the first flight of a steerable airship in the now familiar, and aerodynamically efficient, cigar-shaped "steam balloon."

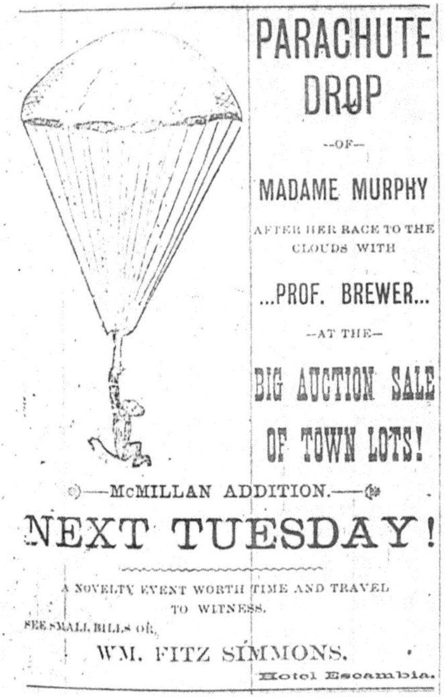

PARACHUTE DROP

--OF--

MADAME MURPHY

AFTER HER RACE TO THE CLOUDS WITH

...PROF. BREWER...

--AT THE--

BIG AUCTION SALE OF TOWN LOTS!

--McMILLAN ADDITION.--

NEXT TUESDAY!

A NOVELTY EVENT WORTH TIME AND TRAVEL TO WITNESS.

SEE SMALL BILLS OR.

WM. FITZ SIMMONS.
Hotel Escambia.

Figure 2. This January 1903 advertisement never quite explains how Madame Murphy ascended to the clouds with Professor Brewer nor why she appears to be a monkey. (Courtesy *Pensacola News Journal*)

FIRST LIGHTER-THAN-AIR FLIGHT OVER PENACOLA

As early as April 2, 1842 the *Pensacola Gazette* reported free balloon activity near Pensacola. Hugh F. Parker, launching from the Public Square in Mobile, was reported to have crash-landed east of the Escambia River. There is no indication, however, that he passed over the city of Pensacola.[5]

A. N. (Tony) Nassr made the first known lighter-than-air flight over Pensacola on November 8, 1910. Invited to fly at the Pensacola Interstate Fair, he flew his steerable, 82-foot long airship from Kupfrian's Park.[6] Once again, a civilian had robbed Naval aviators and Army aviators of their chance to conduct a first flight over Pensacola.

It is very possible that another aeronaut may have made an earlier balloon flight at Pensacola. Balloons were readily available before the turn of the century and captive balloon ascensions were a popular attraction at fairs. In fact, Professor Gowdy provided double captive balloon ascensions at the third Pensacola Interstate Fair in November, 1911.[7] It is not unreasonable to believe that another aeronaut may have conducted either a captive or free balloon flight at Pensacola even earlier than 1910.

U.S. NAVY INTEREST IN BALLOONS

The U.S. Navy's first association with lighter-than-air occurred during the Civil War when balloonists employed by the U.S. Army operated from ships. On August 3, 1861 civilian balloonist John La Mountain made the first ascension from a boat when he rose more than 2,000 feet. He was aboard the Union gunboat FANNY, which was moored near Fort Monroe, at Hampton

Roads, Virginia, to observe confederate activities there.[8] Three months later Professor Thaddeus S.C. Lowe, a civilian aeronaut under contract to the U.S. Army, converted a U.S. Navy coal barge, the USS GEORGE WASHINGTON PARKE CUSTIS, into the world's first dedicated balloon carrier. On November 11, 1861 during the Peninsular Campaign along the Chesapeake Bay, he ascended in his balloon for his first mission from the deck of the USS GEORGE WASHINGTON PARKE CUSTIS after she had been towed along the Potomac River.[9]

The U.S. Navy abandoned any further involvement with lighter-than-air until the early 1900s when Captain Mark Bristol, the Director of Naval Aeronautics, expressed interest in their use in anti-submarine warfare. German airship development, and its subsequent deployment in raids over England during World War I, had generated considerable publicity and interest in military applications by Naval aviators.

Accordingly, in June 1915, almost exactly four years after ordering its first aeroplanes, the U.S. Navy funded the purchase of a kite balloon and free balloon from the Goodyear Company, and a non-rigid airship from the Connecticut Aircraft Company.[10] The logical choice to base all of the U.S. Navy's lighter-than-air assets was the Naval Aeronautic Station at Pensacola.

Included in the purchase of the free and kite balloons were provisions for the lighter-than-air training of naval officers. So in March 1915 Lieutenant Commander Frank R. McCrary was ordered to report to Akron, Ohio for instruction and became the first Naval aviator to qualify as a lighter-than-air pilot.[11] He also established the U.S. Navy's first formal lighter-than-air training program at Pensacola, commanded all of the first free balloon flights at Pensacola, and participated in the trials of the DN-1, the U.S Navy's first airship, at Pensacola.

WOOLSEY RAZED FOR BALLOON FIELD

Planning to accommodate this new training had apparently started several months earlier. Open space on the grounds of the old Pensacola Navy Yard was limited, and since appropriations also called for the construction of a double dirigible hangar and hydrogen plant, additional property was needed to accommodate lighter-than-air activities at Pensacola.[12] The first hint that the U.S. Navy planned to expand aviation activities in Pensacola was given in a September 1914 newspaper article:

Another interesting report is that the navy department is contemplating taking in the whole of Woolsey at the yard for its aerial grounds. This could not be verified yesterday but from the machine which was perceived there yesterday and the fact that other aviators are to arrive shortly it is apparent that there is much ground for the rumor.[13]

Thus, the first expansion beyond the brick walls of the old Pensacola Navy Yard materialized when the U.S. Navy sought a site to accommodate their lighter-than-air training.

Several months passed before the residents of Woolsey were given official notice in April, 1916 to either move or demolish their homes so the land could be cleared for use as a balloon field.[14] While awaiting the completion of construction of the new balloon field at Woolsey, most launches were made from the grounds of the station.

Surprisingly, the razing of Woolsey and the later movement of some Warrington homes were accomplished with very little bitterness when one considers that most of the residents of these two villages had lived there most of their lives.[15]

The Pensacola Pilots' Association, most of whose members lived in Woolsey and Warrington, did prepare a petition in 1916 that requested these razed buildings be appraised and that Congress be approached regarding possible reimbursement for those owners who were displaced.[16]   They were very careful to note that:

It is the purpose of the following resolution not to retard in any way the government in its intention to enlarge the Pensacola navy yard beyond the confines of the present walls, but only to urge that the buildings to be demolished or removed to other parts of the reservation be appraised as to their present standing value and the present owners of such improved property, ... be reimbursed at such appraised valuation. [17]

In 1919, a bill was introduced in the Senate to reimburse those residents of Woolsey and Warrington who were affected by the order to move.[18] It noted the "...apparent hardship to a number of tenants who had been granted permission to reside within the reservation and whose property was removed by order of the department as a necessary incident to the carrying out of the extensive scheme of improvement and enlargement of that station."[19]

It is believed that this bill was never passed.

Figure 3. A kite balloon is prepared for ascent from NAS Pensacola. The first balloon launches were made from inside the station as the village of Woolsey had not yet been cleared. This particular site is in front of Building 16, present home of the Navy Legal Services Office. (Courtesy National Museum of Naval Aviation)

FIRST KITE BALLOON FLIGHT OVER PENACOLA

By April 1, 1916 the first kite balloon had arrived in Pensacola accompanied by Goodyear representatives.[20]   Designed to carry two people, it was described as being "...85 feet in length, made of rubber and somewhat resembled a Zeppelin in shape."[21] The new hydrogen plant was operational and it was only a matter of days before the balloon would make its first flight over Pensacola.

The U.S. Navy's first kite balloon ascent in Pensacola occurred on April 3, 1916 as Goodyear instructors commenced a training course for U.S. Navy personnel in the inflation and handling of the balloon. The flight was clearly visible to local Pensacolians:

For hours yesterday scores of field glasses in the city were trained upon the huge captive 'balloon' which hovered at varying heights over the Pensacola naval station. The latest acquisition of the navy to the local station seemed to have been handled with evident ease, for it was moved about from place to place, attracting consider-able attention.

Attached to the balloon is an appendage, resembling the 'tail' of the kite, and seeming to have the same use, for it appeared to balance the big aerial traveler evenly. A basket, swung many feet below the large gas bag proper, carried several men at times, and this also seemed to have been handled easily upon the first occasion of the try-outs. A representative of the Goodyear Rubber Company, manufacturers, was on hand and was really in charge as demonstrator.[22]

MAN-LIFTING KITES

Before the U.S. Navy experimented with balloons they tested man-lifting kites capable of lifting an observer from the deck of a ship into the air.

One of the early pioneers of these kites was Captain B.F.S. Baden-Powell of the Scots Guards who in 1894 successfully lifted a man on several occasions. In 1897, Lieutenant H.D. Wise of the U.S. Army made several successful kite ascents at Governor's Island.

In January 1911 Samuel Perkins boarded the USS PENNSYLVANIA (ACR-4) to conduct shipboard experiments with man-lifting kites. With the ship underway at 13 knots, two 9-by-9 feet pilot kites attached to

Lieutenant Rodgers rises in a bos'n chair from the deck of the USS PENNSYLVANIA.

several 6-by-6 feet secondary kites, were sent up. These were attached to several 9-by- 9 feet lifting kites connected to a narrow bos'n chair.

The first man to go up was the assistant engineering officer of the USS PENNSYLVANIA, who rose to a height of more than 400 feet. He was Lieutenant John Rodgers who two months later reported to the Wright brothers camp in Dayton Ohio for flight training and eventual designation as Naval Aviator Number 2.

Charlton, A.M. (1938, October). Man-Lifting Kites in the Navy. *Proceedings, 64,* 1452-1455.

KITE BALLOON ESCAPES

Figure 4. A kite balloon is prepared for ascent at NAS Pensacola's balloon field. (Courtesy National Museum of Naval Aviation)

Things really got exciting for Pensacolians a few days later on April 7, 1916. The newspaper told the story:

Shortly after 7 o'clock yesterday morning the balloon was sent up and had reached a height of about eighty feet when a strong puff of wind struck it. The craft immediately began to fight for its release and an effort was made to bring it back to earth. It was anchored to a two-ton anchor firmly imbedded in the ground as well as being held by about thirty guy ropes weighted with sand bags. A cable, used to raise and lower the balloon, was also attached to a large motor truck, the latter being used to regulate the height of the balloon in the air.

All of these combined weights however, were not sufficient and after a struggle of about an hour, the two-ton anchor was pulled from the ground. The driver of the auto truck turned on his engines and endeavored to bring the craft back to earth, and had it down to within thirty feet of the ground, when the guy ropes snapped, breaking the bridle and dropping the cable to the ground.

Away soared the balloon, rising steadily and driven by a thirty mile gale, went at a fast clip in a southeasterly direction. As it passed over the navy yard wall, the six tails struck the brick wall, five of them being torn away. The one remaining tail was not sufficient to maintain the craft in an upright position and it turned over.

Men at the aero station followed the balloon through glasses, but it was lost to sight within thirty minutes and was then proceeding in the direction of Milton.

Had the tails remained intact, experts say the balloon would have remained in the air for an indefinite time, probably months, but this fact and also that it was not fully inflated, caused it to strike the top of a tall pine tree about three miles out from Argyle, where it lodged.[23]

Figure 5. With movies like this playing at the local theater, it is no wonder that people were terrified when a kite balloon resembling a Zeppelin escaped from NAS Pensacola and appeared over their homes without warning. (Courtesy *Pensacola News Journal*)

Argyle is located about 84 miles from Pensacola. With newspapers and movies blazing stories of German Zeppelin's attacking

London some residents believed the "...country had been invaded by Germany and the balloon was none other than a Zeppelin on a raid. The fact that such a balloon was at the navy yard was not generally known and when the eighty-foot monster was seen flying by it caused some apprehension and a great amount of curiosity."[24]

Figure 6. Postcard depicting a balloon inflation within the grounds of the Naval Aeronautic Station. (Author's collection)

The Pensacola Journal reported that the balloon "...was felled by the natives and what was left of the balloon was cut into strips..." so that when Navy rescuers arrive "...they will find only the basket intact."[25] Fortunately, the story had been much exaggerated and the balloon, while missing several large sections, was eventually repaired.[26] In the future, the balloon men would use stone anchors buried in the ground, instead of sand ballast bags, to hold the kite balloon captive "...to ensure against a repetition of the accident." [27]

FIRST FREE BALLOON FLIGHT OVER PENSACOLA

By the end of April 1916 two free balloons also had been received in Pensacola and under the direction of Lieutenant Commander McCrary preparations were made for their first flights.[28] McCrary, and not Goodyear pilots, would conduct nearly all of the first free balloon flights in Pensacola. U.S. Navy lighter-than-air training is considered to have officially started with his first free balloon flight on May 5, 1916.[a]

The Pensacola Journal described one of McCrary's free balloon flights on May 23, 1916:

Closely followed by aeroplanes, and with thousands of people watching its every movement, the navy observation balloon, a great, golden orange flashing

Figure 7. A free balloon prepares to launch from Naval Aeronautic Station Pensacola in front of Building 16, which is still in use today. (Courtesy National Museum of Naval Aviation

[a] However, Goodyear representatives were present who trained both officers and ground crew in the inflation, launching and ground handling of balloons.

in the sunshine, swung over Pensacola and effected a landing on East Hill yesterday afternoon, completing a most successful practice trip.

Promptly at half past two, the big bag left the aeronautical station, Lieut. Commander Frank McCreary (sic) being in charge. Junior Grade Lieutenants W. Gillespie and H. Bartlett also made the flight. A 16-mile wind was blowing, and the balloon took a northeasterly course over Pensacola, the streets of which were thronged with people watching the first successful test of the kind made here.

No difficulty was experienced in landing the balloon. The precision with which this type can be landed is one of its features.

The balloon landed near the corner of Fourteenth Avenue and Eighteenth street and it had hardly settled on the ground before a large crowd of persons, who had been following it, gathered. These included a squad of motorcycle men from the aero station and the big auto truck from that place, into which the balloon was loaded and returned to the station.[29]

McCrary and his free balloons would be a frequent sight over Pensacola and on June 3, 1916 the Secretary of the Navy approved his proposal for the first formal course of instruction in free and kite balloons.[30] On September 17, 1917 the Commandant of Naval Aeronautic Station, Pensacola forwarded to the Bureau of Navigation the names of 10 officers who qualified at Pensacola as free and kite balloon pilots between December 1916 and August 1917.[31]

WOOLSEY BALLOON FIELD

In March 1917 *The Pensacola Journal* described the completed field:

An immense balloon field has been constructed just north of the aeronautic station on the site formerly occupied by the houses of ...[Woolsey]. The field is on the beach extending back for several hundred feet, and is perfectly level.

Figure 8. 1916 view of NAS Pensacola with the villages of Woolsey and Warrington still intact. (Courtesy T.T. Wentworth, Jr Collection, Western Florida Historic Preservation, Inc.)

In the center is a land anchor of about a hundred feet in height surmounted by a large reflector with a light probably for use during the night. It is designed solely for dirigible balloons, and it is probable that when the new double dirigible hangar is constructed it will be placed on this field. [32]

A U.S. Navy report a few years later described improvements to the field:

Some of the dwellings were moved to clear an area upon which have been built a large dirigible balloon hangar, a workshop for dirigible repairs, and four large magazine and mine storage buildings. A floating hangar on a huge pontoon was found impracticable in service and was removed from its pontoon or barge and located near the other dirigible hangar in this Woolsey region. Two brick buildings surrounded by a wall are also in this vicinity, they having been the magazine buildings of the old navy yard. [33]

FREE BALLOON ESCAPES

On June 30, 1917 a second balloon was accidentally released from the station. Chief Petty Officer Gay was working on one of the free balloons when:

The balloon was being adjusted preparatory to a flight. The officer was in the rigging, just between the basket and the bag, making a slight adjustment, when a gust of wind jerked the surging craft from the hands of the helpers and sent it aloft. Instead of attempting to jump as the balloon left the ground, the lone occupant stayed with the vagrant skyship and was blown far out over the bay.

In a short while the balloon had mounted a mile and a half above the earth, with the unwilling passenger hanging halfway between the basket and the bag. It appeared but a speck to the thousand of watchers, who followed breathlessly the efforts of the officer to bring the craft to earth.

Finally he succeeded in slipping down into the basket and pulled the control valve, releasing the gas from the bag. The craft started a sudden descent, and fell into the bay near Santa Rosa Island. Watchers at the station had the high speed boats ready, and arrived at the spot almost as soon as the balloon did. Luckily, Officer Gay was not caught under the bag and was easily rescued, uninjured, after his wild ride. [34]

Figure 9. USS HUNTINGTON (CA-5) receives a kite balloon at Pensacola in June 1917. (Courtesy U.S. Navy)

KITE BALLOONS DURING WORLD WAR I

The balloon was soon replaced and experiments were conducted to determine their suitability to use aboard ship for gunfire spotting and scouting. [35] Testing aboard battleships at Pensacola revealed several significant problems, including the hazards of carrying hydrogen aboard ship, restricted battleship maneuverability when kites were raised, and the poor reliability of the balloons themselves.

However, kite balloons were used extensively by both the U.S. Army and U.S. Navy throughout World War I. More than 642 balloons were manufactured for the U.S. Army and 92 balloon squadrons sent overseas.[36] The U.S. Navy had more than 300 kite and free balloons to combat the German U-boat menace.[37] Towed by ships to 1,000 feet altitudes, and with wired communications between the observer and the ship, they served as convoy escorts and patrols. Kite balloons would see service aboard U.S. Navy ships until July 16, 1922, when the last kite balloon was removed from the USS WRIGHT (AZ-1).[38]

BALLOON RECORDS

Not every record broken at Pensacola involved heavier-than-air. In March 1920, four U.S. Navy balloonists who launched from NAS Pensacola established a distance record for "35,000 cubic foot standard navy balloons" having traveled an air distance of about 700 miles when they landed in Murdock, Illinois.[39]

This record was toppled a month later when another balloon occupied by four Pensacola Naval aviators traveled from Pensacola to Wilmington, North Carolina, a distance of 900 miles.[40]

The U.S. Navy, however, was developing an interest in the German rigid airship program.

LIGHTER-THAN-AIR TERMINOLOGY

The terminology used to describe all the various lighter-than-air vehicles can be somewhat daunting:

1. Free balloons simply float freely with the direction of the wind.

2. Captive balloons remain tethered to the ground and are raised or lowered by means of a winch. These balloons were quite popular at county fairs.

3. Kite balloons are an elongated form of a captive balloon that are fitted with a tail appendage to keep them headed into the wind.

4. Airships differ from balloons in that they are engine-powered and steerable. There are three types:

    a. Non-rigid, the shape is determined by the inflated envelope

    b. Semi-rigid, which contains no framework except for a keel that runs the length of the envelope to which the car is attached

    c. Rigid, the shape is determined by an internal framework of girders

5. Blimps are non-rigid airships.

6. Dirigibles are semi-rigid or rigid airships.

7. Zeppelins are semi-rigid or rigid airships as designed by Count Ferdinand von Zeppelin.

NOTES

[1] Taylor, J.W.R. & Munson, K. (1977). *History of Aviation.* London: Octopus Books, page 20.

[2] Taylor, J.W.R. & Munson, K. (1977). *History of Aviation.* London: Octopus Books, page 20.

[3] Crouch. T.D. (2000). *Flight in America, 1784-1919.* Retrieved from http://crm.cr.nps.gov/archive/23-02/23-02-2.pdf

[4] Clines, C.V. (1996, September). First in America's Skies. *Aviation History,* pages 27-32.

[5] Parker, H.F. (1842, April 2). Balloon Flight. *Pensacola History Illustrated,* Fall 1986, (2) 2, page 11.

[6] A Flight Of Nassr's Ship (1910, November 9). *The Pensacola Journal.*

[7] Wednesday's Program At the Interstate Fair. (1911, November 3). *The Pensacola Journal.*

[8] Layman, R.D. (1979) *To Ascend from a Floating Base.* London: Associated University Presses.

[9] Lindley, J.M. (1978, March). Wings over the Ocean. *Naval Aviation News,* page 37.

[10] Grossnic, R.A. (n.d.) *Kite Balloons to Airships...the Navy's Lighter-than-Air Experience.* Washington, D.C.: Government Printing Office.

[11] Grossnick, R.A. (n.d.) *Kite Balloons to Airships...the Navy's Lighter-than-Air Experience.* Washington, D.C.: Government Printing Office, page 9.

[12] Pearce, G.F. (1980). *The U.S. Navy in Pensacola From Sailing Ships to Naval Aviation (1825-1930).* Pensacola, FL: University Presses of Florida, page 142.

[13] New Aircraft Is Received At Local Station. (1914, September 10). *The Pensacola Journal.*

[14] Residents of Woolsey Are Told To Move. (1916, April 2). *The Pensacola Journal.*

[15] The Air Station A Growing Concern. (1931, July 20). *Air Station News, Pensacola, Florida.*

[16] Enlargement Of The Aviation Plant Is Of First Importance. (1916, April 7). *The Pensacola Journal.*

[17] Enlargement Of The Aviation Plant Is Of First Importance. (1916, April 7). *The Pensacola Journal.*

[18] Warrington People To Be Reimbursed. (1916, August 29). *The Pensacola Journal.*

[19] Relief Of Occupants Of Certain Premises Within The Naval Station At Pensacola, Fla. Report No. 141, 66th Cong. (1919).

[20] An Observation Balloon Reaches The Aero Station. (1916, April 1). *The Pensacola Journal.*

[21] Giant Observation Balloon Escapes From Aero Station. (1916, April 8). *The Pensacola Journal.*

[22] Balloon Is Given A Try-Out At The Aviation Station. (1916, April 4). *The Pensacola Journal.*

[23] Giant Observation Balloon Escapes From Aero Station. (1916, April 8). *The Pensacola Journal.*

[24] Giant Observation Balloon Escapes From Aero Station. (1916, April 8). *The Pensacola Journal.*

[25] Giant Observation Balloon Escapes From Aero Station. (1916, April 8). *The Pensacola Journal.*

[26] Captive Balloon Has Been Repaired. (1916, June 18). *The Pensacola Journal.*

[27] Captive Balloon Has Been Repaired. (1916, June 18). *The Pensacola Journal.*

[28] Will Test Two New Balloons At Aero Station. (1916, April 30). *The Pensacola Journal.*

[29] Balloon Test Scores First Big Success. (1916, May 24). *The Pensacola Journal.*

[30] Grossnick, R.A. (n.d.). *Kite Balloons to Airships...the Navy's Lighter-than-Air Experience.* Washington, D.C.: Government Printing Office, page 9.

[31] Jayne, J.J. (1917, September 17). *Officers Who Qualified As Free And Kite Balloon Pilots.* Pensacola, FL.

[32] Immense Balloon Field Built Here. (1917, March 2). *The Pensacola Journal.*

[33] Bennett, F.M. (1919, March 15). *Historical Information, Collection of.* Special Collections, 73-6, Box 2, John C. Pace Library.

[34] Presence Of Mind Saves Balloon From Loss. (1917, July 1). *The Pensacola Journal.*

[35] Grossnick, R.A. (n.d.) *Kite Balloons to Airships...the Navy's Lighter-than-Air Experience.* Superintendent of Documents Government Printing Office: Washington, DC, page 4.

[36] Fulton, G. (1963). Kite Balloons Afloat. *American Aviation Historical Society,* 8(4), pages 248-253.

[37] Fulton, G. (1963). Kite Balloons Afloat. *American Aviation Historical Society,* 8(4), pages 248-253.

[38] Grossnick, R.A. (n.d.). *Kite Balloons to Airships...the Navy's Lighter-than-Air Experience.* Superintendent of Documents Government Printing Office: Washington, DC, page 22.

[39] Navy Balloon Pilots Return. (1920, March 28). *The Pensacola Journal.*

[40] Balloon Sets New Record. (1920, April 8). *The Pensacola Journal.*

CHAPTER 10
AIRSHIPS AND DIRIGIBLES

Figure 1. May 1920 photograph of the balloon field and dirigible hangars at Naval Air Station Pensacola located north of the old Pensacola Navy Yard on the former site of the village of Woolsey. (Courtesy U.S. Navy)

FIRST AIRSHIPS OVER AMERICA

The first engine-powered airship flight[a] in the United States was made in 1902 by Edward C. Boyce.[1] Flying an imported airship at Brighton Beach, New York, the flight ended prematurely when a maneuvering line fouled his propeller, forcing an emergency landing. The first successful airship in the United States, however, is considered to be the *California Arrow,* constructed by Captain Thomas S. Baldwin and using an engine built by Glenn Curtiss. Captain Baldwin[b] was a veteran aeronaut who made his first balloon ascent in 1875. On August 3, 1904 he made his first powered airship flight in the United States.

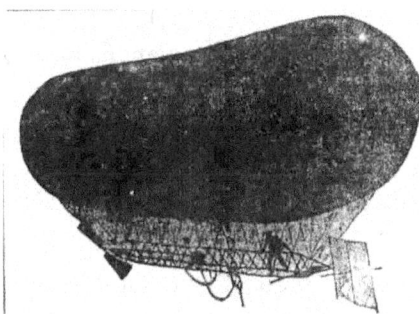

Figure 2. This photograph of Nassr's airship illustrates how the early aeronaut controlled his craft. Walking along a wooden fuselage suspended from the envelope, he shifted his weight fore or aft to counterbalance the weight of the motor and to cause the airship to ascend or descend. A wire leads to the throttle and he used cords to control the steerable rudder aft of the envelope. (Courtesy *Pensacola News Journal*)

FIRST AIRSHIP OVER PENSACOLA

Heralded as the "greatest aerial event ever witnessed in this section of the south" the first airship to fly over Pensacola occurred during the 1910

---

[a] American aeronauts operated pedal-powered airships as early as 1878.

[b] A civilian, Baldwin explained to Glenn Curtiss upon their first meeting that all balloon pilots were called Captain and that he chose to retain this honorific.

Pensacola Interstate Fair.[2] Aeronaut A.N. (Tony) Nassr and his airship arrived at Pensacola's Louisville & Nashville Railway depot on October 25, 1910.[3] A gas making apparatus and an immense tent hangar were constructed at the "air dome" on the fairgrounds at Kupfrian's Park to house the airship, which was described as being 82 feet long and 16 feet in circumference with a seven and one half horse power gasoline engine.[4]

Holder of the United States endurance record at two hours and five minutes, Nassr made his first flight from the fairgrounds on November 8, 1910.[5] The airship was scheduled for two daily flights during the fair and at least one attempt was made to fly over the city of Pensacola before an engine malfunction forced him to abort his attempt. Nassr and his wife remained in Pensacola after the fair to build an aeroplane, occasionally making airship exhibition flights in nearby towns.[6]

DIRIGIBLE, NONRIGID, NO. 1

In June 1915 the U.S. Navy placed an order for its first non-rigid airship, designated the DN-1 (Dirigible, Nonrigid, No. 1). To be constructed by the Connecticut Aircraft Company for $45,636 ($814,929 in 2002 dollars), there was a slight problem in that no one in this company, or in the United States for that matter, had experience building airships. Consequently, although the airship was expected to arrive at Pensacola in October 1915 for initial testing and evaluation, due to manufacturing delays it did not arrive until December 1916.[7]

In preparation for the arrival of the DN-1, Lieutenant Frank R. McCrary was ordered to the Connecticut Aircraft Company in October 1915 to serve as an inspector and receive airship flight instruction.[8]

In April 1916 a separate training division for lighter-than-air was established at Naval Aeronautic Station Pensacola, including pilot and "mechanician" instruction in dirigibles and balloons.[9] Paralleling heavier-than-air training requirements that outstripped Pensacola's capabilities, the demand for lighter-than-air pilots during World War I was so great that additional training sites were established at several Naval air stations across the United States, France and England.

FLOATING DIRIGIBLE HANGAR

Two dirigible hangars were eventually built on the balloon field on the former village of Woolsey to accommodate lighter-than-air activities at Pensacola.

The first was a floating steel and corrugated iron hangar 183' long, 58' wide and 60' tall specifically designed to house the DN-1 at Pensacola. The use of a floating hangar was copied after the German design in which the hangar could be maneuvered "...in the direction of the prevailing wind for ease in airship handling."[10] This hangar was ordered from the American Bridge Company of Pennsylvania with a requirement that it be completed prior to

Figure 3. Inside of the former floating steel hangar after she was reconstructed ashore and used as an aeroplane storage hangar. (Courtesy National Museum of Naval Aviation)

the arrival of the DN-1. Built in two parts, the floating barge part of the hangar was towed from Pittsburgh down the Mississippi River into the Gulf of Mexico and arrived in Pensacola in early 1916.[11] The superstructure was then added and the floating hangar was completed in September 1916.[12] Experiments with removing and storing the DN-1 in this floating hangar proved to be difficult, including the unexpected emergence of a back draft in the rear of the hangar that tended to suck the dirigible in. Also noted was the disadvantage of transporting hydrogen and supplies at sea.[13]

After the demise of the DN-1, this hangar was eventually removed from its floating barge and rebuilt on shore in the July to December 1918 timeframe.[14] Dismantled for sale after suffering severe storm damage on September 28, 1926,[c] it was not moved off station property until 1935.[15]

DN-1 ARRIVES AT PENSACOLA

Tightly packed in crates, the DN-1 arrived in Pensacola on December 14, 1916 aboard a railroad car.[16] The DN-1 stretched more than 175 feet long and was over 50 feet tall. It carried a gondola capable of carrying eight people and was powered by two four-bladed propellers driven by an eight-cylinder

Figure 4. The DN-1 alights on Pensacola Bay on April 27, 1917. (Courtesy National Museum of Naval Aviation)

---

[c] A former worker at the station observed that a lot of the hangar's steel windows were purchased and installed in several Warrington homes and businesses.

Figure 5. The DN-1's floating hangar moored inside the boat basin at NAS Pensacola. Note the airship in the background. (Courtesy National Museum of Naval Aviation)

motor.[17] Captain Thomas S. Baldwin, now chief constructor for the Connecticut Aircraft Company, described her as having a "...coat of bright orange varnish...found to repel the actinic blue and violet rays of the sun, which burn up the inner coating of rubber."[18]

On April 20, 1917 the DN-1 made her inaugural voyage with the Connecticut Aircraft Company's pilot, Hans O. Stagel, and crew aboard. *The Pensacola Journal* and Mr. Smyth, General Manager of the Connecticut Aircraft Company, heralded the success of the DN-1 the next day:

Leaving the floating hangar at the aeronautic station at 5 o'clock Friday afternoon, the DN-1 ascended to a thousand foot altitude, headed across the bay, sailed over Santa Rosa Island, then pointed shoreward, circled over Pensacola, and returned to the yard – the first voyage of American's first dirigible, and a perfect success.[19]

Despite the bright face put on by the U.S. Navy for an admiring public, after only two more flights the DN-1 was ignobly "...put back in its shed, deflated, and eventually broken up."[20]

Design problems, lack of U.S. dirigible manufacturing and engineering expertise, and material problems doomed this poor airship from the outset. One author noted:

Actually the craft was a failure, proving to be far too heavy to lift itself off the ground. In addition the envelope leaked badly. Contractor and Navy alike despaired of ever getting the contraption aloft. In a desperate attempt to reduce weight, one engine was removed and a sprocket and chain drive to the two propellers substituted for the gear and shaft arrangement. Even so, the craft could barely get aloft. Although the acceptance tests were unsatisfactory it was decided that the contractor had acted in good faith and the contract price was paid.[21]

Later known as the A-class, the DN-1 was the only airship of her type built.

SECOND DIRIGIBLE HANGAR

The Connecticut Aircraft Company also was constructing two other dirigibles for the U.S. Navy. In anticipation of their arrival a second steel dirigible hangar was built at Pensacola that was 250' long, 128' wide and 97' tall. Construction of this hangar began in March 1917 and by October 1917 it was completed.[22] This hangar was not dismantled until 1954.

Figure 6. A B-class airship, follow-on to the DN-1 , on a flight near NAS Pensacola. (Courtesy National Museum of Naval Aviation)

GOODYEAR AIRSHIP TRAINING

When contracts were awarded for the B-class airship (the follow on to the DN-1), Goodyear proposed to train lighter-than-air pilots since there was no U.S. Navy program. The U.S. Navy accepted and on May 29, 1917 an agreement was reached for the establishment of a Naval aviation training detachment at Wingfoot Lake near Akron, Ohio.

On September 26, 1917 with Lieutenant Lewis H. Maxfield assigned as officer-in-charge, the first students started training under the guidance of Goodyear instructors.[23] Training consisted of ground school, flights in kite balloons and free balloons, and finally qualification flights in the B-class airship.[24] With World War I raging, graduates were immediately sent overseas or to lighter-than-air patrol and training stations along the Atlantic coast.[25] Eventually, as more Navy pilots were qualified, the U.S. Navy assumed responsibility for this training facility and operated it as a U.S. Naval Airship Training Station from 1917 through 1921.[26]

Figure 7. There appear to be few ground handlers maneuvering a B-Class airship in front of NAS Pensacola's large dirigible hangar in 1918. (Courtesy National Museum of Naval Aviation)

B-CLASS AIRSHIP ARRIVES AT PENSACOLA

The B-class dirigible proved a much greater success than the DN-1. Design specifications called for a top speed of 45 miles per hour, a 12-hour endurance at 35 miles per hour, and the capability of landing at sea and for being towed.[27] All of these parameters were exceeded. The B-1 was shipped to Pensacola on August 7, 1917 and its first flight was on August 29.[28] Sixteen B-class airships were eventually purchased and while none were shipped to Europe for service during World War I, the U.S. Navy used them extensively for coastal patrol and training. [29]

Figure 8. Close up of the B-class airship control car. Described as "little more than an airplane fuselage suspended beneath an envelope" it held a pilot, co-pilot and engineer. (Courtesy National Museum of Naval Aviation)

C-CLASS AIRSHIP ARRIVES AT PENSACOLA

The next airship in the U.S. Navy series was the C-class, designed in 1918 with increased engine power, endurance, and payload for World War I anti-submarine warfare patrol and convoy protection duties. In April 1920, *The Pensacola Journal* announced that the C-7 soon would be put in commission at NAS Pensacola.[30] While only 10 eventually were built due to the end of World War I, this was an airship of many firsts:

Figure 9. C-class airship C-7 in front of the dirigible hangar at NAS Pensacola. (Courtesy U.S. Navy)

First to carry an airplane aloft and launch it in flight; to demonstrate the practicality of aerial refueling from ships at sea by taking on fuel from a submarine chaser; to use helium as her lifting gas; to complete a transcontinental flight across the U.S.; and to make numerous record-setting distance flights.[31]

D-CLASS DIRIGIBLES

The D-class dirigible, of which five were constructed beginning in 1920, was notable in that for the first time, an airship of a new class would not be sent to Pensacola. They would also not last long in the U.S. Navy. One was destroyed in a fire and the rest eventually transferred to the U.S. Army.

E-CLASS DIRIGIBLES

A single E-class airship, designated the E-1, was purchased and shipped to Pensacola on December 16, 1918. She would spend her entire career here until stricken from inventory on September 5, 1924.[32]

Figure 10.   The E-1 airship at NAS Pensacola.   (Courtesy National Museum of Naval Aviation)

The U.S. Navy constructed several additional classes of non-rigid airships but Naval aviation experts were monitoring closely German Zeppelin operations during World War I. The U.S. Navy was very interested in rigid airships.

THE GOOD GOAT 'BLIMP'

After a forced landing by the E-1 on the Bayou Texar shoreline in February 1919, the aeronauts noticed a young goat that showed a particular interest in the goings on.   Dubbed 'Blimp,' the goat returned to the Naval Air Station as a mascot for the E-1.

In a rather sad story, on March 15, 1919 *The Pensacola Journal* noted the passing on of Blimp.   Apparently Blimp ate a piece of fabric dope and the newspaper noted that "the effects caused its death. But only a goat would eat dope, and his death was not in the line of duty."

Figure 11.   The large dirigible hangar built on the station's balloon field.   Note homes from the former village of Woolsey to the left of the hangar. (Courtesy National Museum of Naval Aviation)

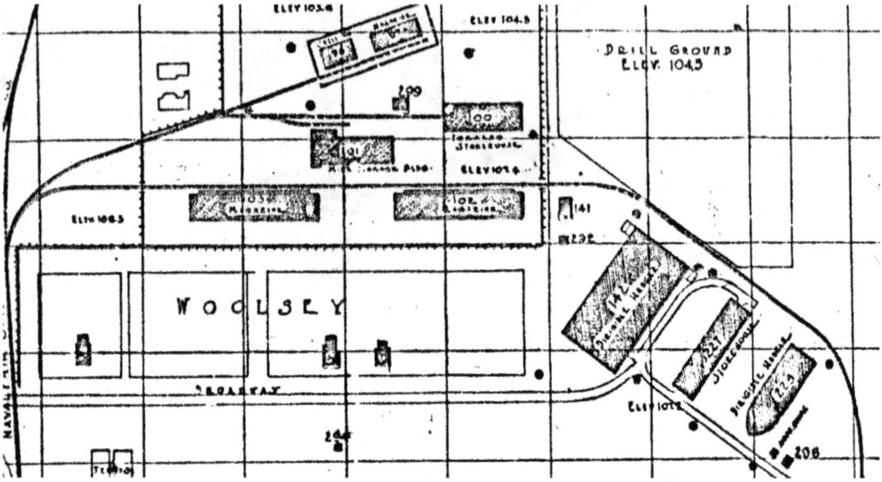

Figure 12. Detail from a July 1, 1921 NAS Pensacola map showing what appears to be three surviving homes from the village of Woolsey to the left with the two dirigible hangars in place to the lower right. (Courtesy U.S. Navy)

ARTHUR E. FORSTER AND AIRSHIP LANDING GEAR

The first airships that arrived at Pensacola were not equipped with landing gear and the control car was simply set down upon the ground. It soon became apparent that the unavoidable bouncing and dragging of an airship landing would cause damage.

Arthur E. Forster, a civilian mechanic at the station, was tasked with fabricating a landing gear. Improvising, he used the rubber flotation bags that were sent with each airship. About 18 inches in diameter and four feet in length, he devised a way to attach them to the bottom of each control car. They were so successful that he was asked to make dozens more and every airship at the station was equipped with them from that point on.

Forster continued to work at Naval Air Station Pensacola until 1955 and retired as Master Mechanic of the Overhaul and Repair Department. An avid historian, many of his photographs and stories are available at the Pensacola Historical Society. Building 3644 at the Naval Aviation Technical Training Center at NAS Pensacola is named in his honor.

PENSACOLA NAVAL AIRSHIP SCHOOL

In January 1919 it was announced at Naval Air Station Pensacola that the "dirigible section has been transferred from the aviation school, and converted into a separate unit, having as the commanding officer Lieutenant (Junior Grade) H.R. Geddee who formerly was senior flight officer of the dirigible section..."[33] The free balloon, observation and kite balloon divisions also were transferred to the Naval Airship School.[34]

Apparently, Geddee did not want his aviators to be daytime flyers only as *The Pensacola Journal* reported that "night flying was one of the first changes of the new administration of the Naval Airship School."[35] In February 1919, experiments also were made at Pensacola with launching torpedoes from a dirigible.[36]

In October 1919 it was announced that "…all naval balloon schools in the east will be consolidated at Pensacola and the air station at

Figure 13. The balloon field at NAS Pensacola. (Courtesy National Museum of Naval Aviation)

that place will be made the permanent training point for all naval aeronauts."[37] Additionally, a local boy, Lieutenant John G. Gibson, son of Mr. and Mrs. J. J. Gibson of Pensacola, was selected to head the balloon school, which included both elementary and advanced lighter-than-air training.[38]

U.S. NAVY RIGID AIRSHIP PROGRAM

In July 1919, the U.S. Navy received approval for the procurement of rigid airships and the establishment of a Naval air station dedicated to lighter-than-air activities.[39] The site selected for the U.S. Navy's first rigid airship base was Lakehurst, New Jersey, which would later absorb Pensacola's entire lighter-than-air training program.

The U.S. Navy eventually contracted for five rigid airships, but only one would visit Pensacola during the timeframe of this book, the USS LOS ANGELES (ZR-3).

USS LOS ANGELES VISITS PENSACOLA

Just as birds are harbingers of spring, so was the arrival of the USS PATOKA (AV-6) a sign that a U.S. Navy rigid airship was operating nearby. A converted oiler, she was the U.S. Navy's only airship tender and she provided a floating mooring mast, helium, and workshops for repairs and maintenance.

On April 22, 1927 the USS PATOKA anchored in Pensacola

Figure 14. The USS LOS ANGELES moored to the USS PATOKA. (Courtesy National Museum of Naval Aviation)

Bay off the Naval Air Station and waited for the arrival of the USS LOS ANGELES.[40] As described by *The Pensacola Journal* on April 25, 1927:

The Los Angeles was sighted a few minutes before 3 o'clock when she appeared on the eastern horizon above Santa Rosa Sound. As she sailed her course up the sound, one of the agile heavier-than-air hydroplanes from the local station, was seen playing about the massive gas bag much as a sparrow teases a large bird. The little ship had gone down the sound early in the afternoon to meet its big sister and accompany her to port.[41]

Several hours later the USS LOS ANGELES gently moored to the USS PATOKA.

ILL-FATED AIRSHIPS

Of the five rigid airships in the U.S. Navy's lighter-than-air program, all but one was destroyed in a crash:

- The USS SHENANDOAH (ZR-1) crashed in Ohio on September 2, 1925 killing 14 of the 43 crewmen aboard.

- The ZR-2 , under construction by the British and as yet unnamed, crashed on August 24, 1922 in England killing 44 of the 49 crewman aboard, including several U.S. Navy personnel.

- The USS LOS ANGELES (ZR-3) was decommissioned on June 30, 1932. Dismantled at NAS Lakehurst, New Jersey in 1939, no other rigid airship in the world served longer.

- The USS AKRON (ZR-4) crashed on April 3, 1933 in the Atlantic Ocean, killing all 73 crewmen aboard.

- The USS MACON (ZR-5) crashed on January 12, 1935 killing two of the 83 crewmen aboard.

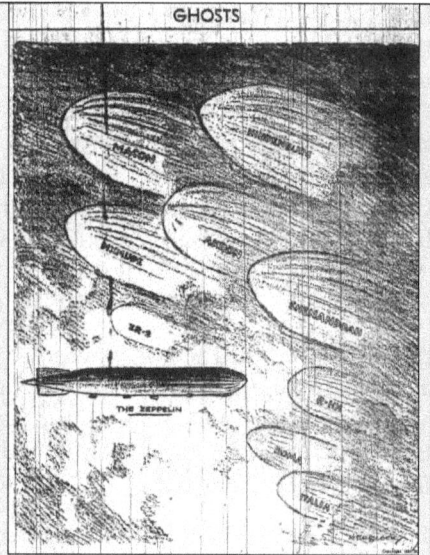

GHOSTS

VISIT OF U.S. ARMY RIGID AIRSHIPS

The U.S. Army purchased its first dirigible in 1908 and operated it until 1912. While interest in lighter-than-air waned during the next several years, World War I revived the U.S. Army's interest and more than 89 balloon companies were organized.[42] In 1920, a joint Army-Navy board decision assigned semi-rigid and non-rigid airship development to the U.S. Army and rigid airship development to the U.S. Navy.[43] Consequently, several airships were transferred from the U.S. Navy to the U.S. Army although both services continued to operate balloons and airships for individual training requirements. The U.S. Army operated their airships from several different bases around the country.

On June 27, 1927 the U.S. Army dirigible TC-5 arrived at the Pensacola Naval Air Station for repairs to her tail. Stationed at Langley Field in Virginia she was operating at Fort Benning on a photographic mission when she was damaged. The airship was flown to Pensacola for repairs, as it was "the only station in the south with hangars large enough to accommodate the craft."[44]

LOSS OF THE C-5

In 1919 there was an informal competition between the U.S. Navy's heavier-than-air and lighter-than-air branches to become the first to cross the Atlantic Ocean by air.

On May 8, 1919 three huge Navy-Curtiss (NC) flying boats departed NAS Rockaway Beach, New York for Trepassey Bay, Newfoundland, Canada on the first leg of their transatlantic voyage. With not enough range to fly non-stop between Newfoundland and Ireland, the shortest distance between North America and Europe, these crews would refuel in the Azores before continuing on to Portugal.

On the same day, the C-class airship C-5, departed NAS Cape May, New Jersey on the U.S. Navy's second transatlantic attempt. Unlike the NC boats, however, the C-5 did have the range to fly non-stop between Canada and Ireland. "Beat the NCs!" became the order of the day for the aeronauts. After a stop at NAS Montauk, New York, the C-5 arrived at her "jump-off point," St. John's, Newfoundland on May 15, 1919. The airship was tied down securely but a few hours later a powerful gale tore the craft from her steel moorings. Unmanned, the C-5 was blown out to sea. Briefly sighted by a British freighter 300 miles northeast of St. John's, she was never seen again.

The next evening the NC boats departed for the Azores and on May 27, 1919, the NC-4 landed in Lisbon, Portugal's harbor. The first transatlantic aviation flight in history had been accomplished.

The NC-4 visited NAS Pensacola at least twice after her historic flight.

DIRIGIBLE ACCIDENTS

Proving that even dirigibles are not immune to engine-out emergency landings, *The Pensacola Journal* reported the following story in February 1919:

Ensign Bruce of the Naval Airship School, crack pilot of dirigible balloons, demonstrated yesterday evening that he was master of the situation when the Naval Dirigible, E-1, of which he was pilot, met with motor trouble and a free balloon landing had to be made.

The accident occurred on the east shore of Bayou Texar and the landing was made successfully. It was stated that the connecting rod in the air pump attached to the huge motor, broke, causing the engine to die.

The huge airship had lost much of the gas which it is inflated due to contraction and a free balloon landing had to be made. Although the crew secured the greatest lift possible, the airship nearly landed in the bayou, and in order to gain additional lift with which to reach land Ensign Bruce ordered the storage battery, with which the airship was equipped, to be cast overboard.

Released of this weight the huge gas envelope soared again and landing on the shore of the Bayou was effected. The only damage sustained was to the flippers, and rudder, which caught in the trees.[45]

LIGHTER-THAN-AIR DISCONTINUED AT PENSACOLA

Unfortunately, the days (and nights) of lighter-than-air training at Pensacola were coming to an end. Based upon the recommendation of Commander Lewis H. Maxfield, on May 16, 1919 funds were authorized for the purchase of a former U.S. Army camp at Lakehurst, New Jersey for use as a U.S. Navy's first rigid airship station.[46]   A huge hangar 943' long, 350' high and 200' wide was constructed and a large landing field cleared and graded."[47]

Commissioned as Naval Air Station Lakehurst in August 1921 all lighter-than-air activities, including balloon, non-rigid airship and rigid airship training were stopped at Pensacola. After a temporary relocation to Naval Air Station Hampton Roads, Virginia, all lighter-than-air training was permanently moved to Naval Air Station Lakehurst.[48] It was the end of an era for Pensacola.[d]

However, Pensacola's proud airship hangars did not sit empty for long. In addition to housing aeroplanes, they also were used for other functions. A November 1921 newspaper article noted that "...since the lighter than air department has been discontinued at the Naval Air Station one of the giant dirigible hangars has been overhauled and fitted up as an amusement auditorium...."[49] Insult had been added to injury.[e]

---

[d] Of interest, the U.S. Navy is considering the reintroduction of lighter-than-air dirigibles and seaplanes to accommodate a need for a fast intermediate logistics capacity.

[e] The last flight of a U.S. Navy airship was at NAS Lakehurst on August 31, 1962.

NOTES

[1] Crouch, T. (1977). The Gasbag Era. *Aviation Quarterly, 3(4),* 290-301.

[2] Nassr's Aerial Wonder is Here for Big Fair. (1910, October 27). *The Pensacola Journal.*

[3] Air Ship Came To Pensacola Last Evening. (1910, October 25). *The Pensacola Journal.*

[4] Nassr's Aerial Wonder is Here for Big Fair. (1910, October 27). *The Pensacola Journal.*

[5] Airship to Fly Today at Big Fair. (1910, November 8). *The Pensacola Journal.*

[6] Will Make Flight In A Dirigible. (1911, January 24). *The Pensacola Journal.*

[7] Rankin, R.H. (1958). DN-1, The U.S. Navy's First Dirigible. *American Aviation Historical Society Journal.* 3(1), page 9.

[8] Grossnick, R.A. (n.d.). *Kite Balloons to Airships...the Navy's Lighter-than-Air Experience.* Superintendent of Documents Government Printing Office: Washington, DC, page 10.

[9] Bennett, F.M. (1919, March 15*). Historical Data, Collection of.* Special Collections, 73-6, Box 2, John C. Pace Library.

[10] Shock, J.R. (1996) *American Airship Bases & Facilities.* New Smyrna Beach, FL: M & T Printers, page 17.

[11] Big Floating Hangar Ready To Come Here. (1916, March 25). *The Pensacola Journal.*

[12] Floating Dirigible Hangar Complete. (1916, September 28). *The Pensacola Journal.*

[13] McCrary, F.R. (1917, August 30). *Memorandum to Commandant,* Arthur E. Forster Papers, Pensacola Historical Society.

[14] Bennett, F.M. (1919, March 15). *Historical Data, Collection of.* Special Collections, 73-6, Box 2, John C. Pace Library.

[15] Shock, J.R. (1996) *American Airship Bases & Facilities.* New Smyrna Beach, Fl: M & T Printers, page 17.

[16] Dirigible Balloon For Aero Station Reaches Pensacola. (1916, December 15). *The Pensacola Journal.*

[17] Dirigible Balloon For Aero Station Reaches Pensacola. (1916, December 15). *The Pensacola Journal.*

[18] Navy's New Pet, A Dirigible, To Have Silk Coat. (1915, June 7). *The Pensacola Journal.*

[19] Maiden Voyage of Navy's First Dirigible Yesterday Exceeds All Expectations. (1917, April 21). *The Pensacola Journal.*

[20] Althoff, W.F. (1990). *Skyships.* New York: Orion Books, page 4.

[21] Rankin, R.H. (1958). DN-1, The U.S. Navy's First Dirigible. *American Aviation Historical Society Journal.* 3(1), page 9.

[22] Navy Buildings Near Completion. (1917, October 30). *The Pensacola Journal.*

[23] Shock, J.R. (1996) *American Airship Bases & Facilities.* New Smyrna Beach, Fl: M & T Printers, page 13.

[24] Grossnick, R.A. (n.d.). *Kite Balloons to Airships...the Navy's Lighter-than-Air Experience.* Superintendent of Documents Government Printing Office: Washington, DC, page 10.

[25] Van Wyen, A.L. (1969). *Naval Aviation in World War I.* Washington, DC: U.S. Government Printing Office.

[26] A Brief History of the Wingfoot Lake Airship Base. (n.d.) Retrieved from http://www.goodyearblimp.com/wingfoot.html.

[27] Grossnick, R.A. (n.d.). *Kite Balloons to Airships...the Navy's Lighter-than-Air Experience.* Superintendent of Documents Government Printing Office: Washington, DC, page 6.

[28] Good Flight Navy 'Blimp'. (1917, August 30). *The Pensacola Journal.*

[29] Grossnick, R.A. (n.d.). *Kite Balloons to Airships...the Navy's Lighter-than-Air Experience.* Superintendent of Documents Government Printing Office: Washington, DC, page 8.

[30] Navy's Greatest Dirigible Almost Ready For Service At Pensacola Naval Air Station. (1920, April 3). *The Pensacola Journal.*

[31] Grossnick, R.A. (n.d) *Kite Balloons to Airships...the Navy's Lighter-than-Air Experience.* Superintendent of Documents Government Printing Office: Washington, DC, page 18.

[32] Grossnick, R.A. (n.d.). *Kite Balloons to Airships...the Navy's Lighter-than-Air Experience.* Superintendent of Documents Government Printing Office: Washington, DC, page 20.

[33] New School Is Formed At Air Station. (1919, January 17). *The Pensacola Journal.*

[34] New School Is Formed At Air Station. (1919, January 17). *The Pensacola Journal.*

[35] New School Is Formed At Air Station. (1919, January 17). *The Pensacola Journal.*

[36] Aerial Torpedo Launched From Blimp At Post. (1919, February 7). *The Pensacola Journal.*

[37] Navy Balloon School To Be Founded Here. (1919, October 10). *The Pensacola Journal.*

[38] Training Base Balloon Pilots Permanent One. (1919, October 21). *The Pensacola Journal.*

[39] Shock, J.R. (1996) *American Airship Bases & Facilities.* New Smyrna Beach, FL: M & T Printers, page 3.

[40] Los Angeles Visits Station. (1927, May 5). *Air Station News, Pensacola, Florida.*

[41] Giant Dirigible Arrives in City For Brief Stay. (1927, April 25). *The Pensacola Journal.*

[42] Mac Closkey, M. (1968). *From Gasbags to Spaceships.* New York: Richards Rosen Press, page 91.

[43] Shock, J.R. (1996). *American Airship Bases and Facilities.* Edgewater, FL: Atlantic Productions.

[44] Army Airship Is Here For Repairs. (1927, June 28). *The Pensacola Journal.*

[45] Blimp Falls On Shores Of Bayou Texar. (1919, February 18). *The Pensacola Journal.*

[46] Althoff, W. F. (1990). *Sky Ships.* New York: Orion Books, page 11.

[47] Althoff, W. F. (1990). *Sky Ships.* New York: Orion Books, page 11.

[48] U.S. Naval Air Station Will Celebrate Centennial Soon. (1925, November 8). *The Pensacola Journal.*

[49] Present Program Monday At Station. (1921, November 20). *The Pensacola Journal.*

CHAPTER 11
WORLD WAR I

Figure 1. Superb photograph that illustrates the World War I expansion of NAS Pensacola along Warrington's waterfront. The village's homes and businesses are visible to the north of the new wooden and steel hangars built to accommodate the increased number of seaplanes assigned to the station. (Courtesy National Museum of Naval Aviation)

WARTIME CHALLENGES FOR PENSACOLA

For the citizens of Pensacola and the aviators of Naval Aeronautic Station Pensacola, World War I presented three great challenges. The first was the physical accommodation of the enormous increase in aeroplanes, students and staff at the station. The second was the arrival of several competitors for Naval aviation resources with the construction of dozens of air stations across the United States, robbing Pensacola of her status as the U.S. Navy's only aeronautic station. The third challenge, and greatest threat, was the wartime introduction of landplanes into the Naval service. If the latter was not addressed properly by the civilian officials of Pensacola, there was a very real danger that Pensacola would lose her solitary hold on initial Naval aviation training. Worst, the aeronautical station might cease to exist altogether.

WARTIME PREPARATIONS

When World War I erupted in Europe in July 1914, the entry of the United States into the conflict was inevitable, yet many powerful Americans hoped to remain neutral and Congressional funding reflected this fact. Despite the fact that foreign countries were spending millions of dollars in developing their own air forces, support for the establishment of a strong Naval air service was slow in developing in the U.S. Navy. Consequently, Pensacola's Naval aviators lived a hand-to-mouth existence with few aeroplanes, fewer parts, and even less capital.

Yet despite this austere funding, in May 1914 Pensacola's Senator and Congressmen felt the outlook was bright for their aviation station.[1] They were particularly encouraged by a statement from the Secretary of the Navy that "...the aviation station is not a temporary affair, but permanent and will be rapidly improved."[2] They hoped to get several construction projects approved by Congress, including construction of three dirigible sheds, 30 permanent hangars, and machine shops for aeroplane repair and part manufacturing.[3]

Precisely where the money would come from to pay for these items was never clearly defined and was probably not available. Naval aviation funding had been so inadequate and poorly utilized, that by the end of 1914 the General Board[a] declared the status of aeronautics in the U.S. Navy as "...nothing less than deplorable."[4]

However, this would soon change when Congress finally began to appropriate the huge sums of money necessary to prepare the United States for entry into World War I. Responding to criticisms of a lack of preparedness, and reacting to the successes of aeroplanes in combat over Europe, the Naval Appropriations Act of March 3, 1915 authorized $1 million for Naval aviation. This was the first specific appropriation for Naval aviation in its short history.[5]

With Naval aviation firmly established here, the timing was perfect for Pensacola. As the only aeronautic station in the U.S. Navy, every dollar authorized meant more money was available locally to improve the station's infrastructure and prepare it for the certain increase in wartime training requirements. The financial floodgates finally were opened, and in less than three years $67 million were appropriated for Naval aviation's use.[6] Pensacola's businessmen were invited to take a share of this largess.

In August 1915, the station was authorized to spend immediately $250,000 for hangar construction, building remodeling and the reopening of machine shops that might employ up to 25 skilled workers.[7] By the end of the year, nearly all of the $1 million appropriation had been spent. Of this, $771,000 was expended on ordering "...twenty aeroplanes, seventy-three aeronautic motors, one free balloon, one floating dirigible shed, one hydrogen plant for dirigibles, one set aeroplane hangars, and one aeroplane wrecking derrick."[8] An additional $130,000 was allocated for repairs and operating expenses at Pensacola.

---

[a] Comprised of senior Naval officers, the General Board served as advisors to the Secretary of the Navy. The position of Chief of Naval Operations had not been established and the current president was Admiral of the Navy George Dewey, the "Hero of Manila Bay" and a strong aviation advocate.

NEED FOR HANGARS

The station soon was flooded with new aeroplanes. As a former shipyard, the facilities at the Pensacola Navy Yard were certainly not designed to meet the unique needs of aeronautical activities. Special facilities were needed for aeroplane assembly, repair and maintenance.

Figure 2. Hangar 73, one of two surviving hangars at NAS Pensacola from the 1916 hangar construction boom. (Courtesy of author)

More importantly, the fragile fabric and wood aeroplanes of this era needed far better protection from the elements than the old canvas tents provided. Accordingly, a lack of permanent hangars was identified as the station's chief problem.[9] The situation became so bad that as new aeroplanes arrived at the Naval Aeronautic Station some could not even be unpacked due to a lack of suitable hangar space. It was "...not considered a good practice to allow the machines to remain crated any longer than is absolutely necessary as the wings are damaged."[10] As one person noted:

This station was built for a Navy Yard, and consequently many of its structures are suited to aviation activities only to the extent of being equipped with four walls and a roof – otherwise they are about as well adapted to the economic and efficient operation of an air station as a Belgian draft horse would be to the Kentucky Derby.[11]

The U.S. Navy needed a permanent hangar building program in Pensacola and one was soon initiated.[12] In October 1915, bids were opened for the construction of three steel frame hangars to be located on the Naval Aeronautic Station's grounds adjacent to the original flying beach. Replacing the original tent hangars still in use, this represented the "first permanent improvements made there in years."[13] The hangars were built several feet above beach level, necessitating the construction of concrete seawalls and concrete seaplane ramps.[14] Gone were the wooden runways traversing the sand into the water. However, it soon became clear that even these hangars could not accommodate all the aeroplanes arriving on station. Additional space, and waterfront, was needed.

PARTIAL MOVE OF WARRINGTON

With the station flying beach fully occupied, the U.S. Navy looked to the west to expand. Since only waterfront property was needed for hangars and seaplane ramps, only those Warrington homes and businesses that bordered the bay west of the station walls needed to be removed. The remainder of the town could stay.[b]

---

[b] On July 10, 1930 the U.S. Navy confirmed reports that additional expansion of the Naval Air Station would require all of the land of the village of Warrington. Between August 7, 1930 and September 15, 1931, 314 buildings were either razed or relocated. Many of the residents moved their homes and businesses across Big Bayou to the present location of Warrington outside the station's front gate.

In August 1916, *The Pensacola Journal* noted that "many of the dwellings that have been moved are now facing north along the car track, the row running to the entrance of the reservation. In many cases where the sand hills have made a slight obstruction some difficulty was met, and residences were cut into two parts to facilitate moving."[15]

Within a month, nearly all the buildings were moved and preparations completed for the construction of three additional steel frame seaplane hangars along Warrington's waterfront. A newspaper article described the hangars as "...built of steel framework, and covered with corrugated metal, having ample lighting facilities through large glass windows in the sides and ends."[16] However, the station's demand for hangar space would prove insatiable, and within a year, the old canvas hangars were set up again around the station and the construction of dozens of temporary wooden hangars completed.[17]

Figure 3. The only remaining building from the villages of Warrington and Woolsey is this former grocery store currently being used as the Navy's Public Affairs Office. (Courtesy of author)

PENSACOLA NAVY YARD EXCEEDED

A major milestone was reached in 1916 when an editorial in *The Pensacola Journal* noted that the payroll of the Naval Aeronautic Station now exceeded that of the old Pensacola Navy Yard.[18] Yet, some Pensacolians refused to give up the idea that the U.S. Navy might reopen an industrial shipyard here, no matter how many times they were rebuffed by Naval authorities. However, for the first time it became evident that perhaps Naval Aeronautic Station Pensacola might prove to be of greater economic value, and more permanent, than a shipyard. This bright future was assured a year later when a Naval commission report recommended the continued development of Pensacola as an aeronautical training and experimental station.[19]

FLURRY OF ACTIVITY

In the fall of 1916 Naval Aeronautic Station Pensacola presented a flurry of activity to a visiting reporter. He noted the construction of three new hangars, shops assembling and rebuilding flying machines, construction of the DN-1's floating hangar, testing of free and kite balloons on the aviation field, and removal of houses from Warrington and Woolsey.[20] He mentioned that 30 additional flying machines were recently ordered and that "...indications are that the station will exceed the activities of the navy yard in its palmiest days."[21] What he didn't realize was that in spite of all this activity, the current capacity for training pilots at Pensacola was a meager 40 pilots per year.

WAR DECLARED

Whatever hopes the United States had of remaining neutral disappeared in January 1917 when Germany declared unrestricted submarine warfare against any ship, regardless of nationality, that entered the war zone. When the United States declared war on Germany on April 6, 1917, both the U.S. Army's and U.S. Navy's aviation arms were unprepared for combat in Europe.

Naval aviation's total assets were 48 officers, 239 enlisted men, and 54 aeroplanes.[22] U.S. Army aviation was larger, with 131 officers, 1,087 enlisted men and 250 aeroplanes.[23] However, neither service's aeroplanes were suitable for wartime operations in Europe nor was there a single aeroplane manufacturer in the country capable of designing and producing a combat aeroplane.

An early Naval aviator described their typical aeroplane:

The standard trainer was a Curtiss AH single float, two-place, pusher-type seaplane powered by a Curtiss OX engine developing about 90 hp, practically all of which was needed to maintain the airplane in horizontal flight at an air speed of from 60 to 65 mph. It's aerodynamic efficiency was impaired by a lack of stream lining and by the full exposure of its occupants to the air stream. Tail surfaces were supported by bamboo outriggers. Naval aircraft therefore, of the period immediately prior to the entrance of this country in the World War, may be said to have been mere contraptions of wood and wire. They had no offensive potentialities, and provided only a perch from which aerial observations could be conducted.[24]

Worse, the United States aviation industry as a whole was in disarray. There were few airplane factories, aeronautical engineers, and workers skilled in producing and maintaining aircraft. Commercial aviation was virtually nonexistent and there were only a handful of airfields across the country. Although the first country to fly in the world, the United States had fallen so far behind the Europeans in aeronautics that our first military aviators were sent overseas without any aeroplanes, forced to borrow some from our allies. More than a year would pass before the first wartime mission was flown "by American pilots, in American planes, powered by American engines."[25] Consequently, World War I served as a major impetus for the development of American aviation in aeroplane design, production and both commercial and military operations.

A joint Army-Navy technical board was formed in May 1917 to coordinate efforts between the services. Oversight was provided on logging operations to ensure that adequate stocks of spruce were available for aeroplane construction, enough fabric was produced for wing and fuselage coverage, and the best aeroplanes were selected for manufacturing.

A British design, the De Havilland DH-4, was selected for landplane production. More than 4,800 eventually were produced with the first American built DH-4's arriving in France in May 1918.[26] A Curtiss design, initially the single engine HS-2 and later the twin engine H-16, was selected for seaplane production.[27] With additional redesigned models more than 1,600 seaplanes were produced to serve from coastal bases.

These were all powered by the brilliant American designed Liberty engine. Nearly 21,000 Liberty engines were built for the United States and her allies before the war ended.

Figure 4. U.S. Navy recruiting poster emphasizing the fact that U.S. Naval aviation was the "first American armed forces to set foot in Europe" during World War I. Most of these aviators were from Pensacola. (Author's collection)

FIRST AMERICAN FORCES IN EUROPE

Almost immediately after war was declared, in response to a request by the French government for the deployment of naval forces to France, Pensacola was ordered by the Navy Department to form an aviation detachment for duties overseas. The First Aeronautic Detachment, under the command of Lieutenant Kenneth Whiting and comprised of seven officers and 122 enlisted men mostly from Pensacola, departed for France aboard two Navy colliers, the USS JUPITER and USS NEPTUNE.[28]

On June 5, 1917 the USS JUPITER[c] arrived at Pauillac, France and on June 8, 1917 the USS NEPTUNE arrived at St. Nazaire, France. The detachment quickly was divided between the British, French and Italians for flight training and equipment with aeroplanes. This Naval aviation detachment was the first American armed forces to arrive in Europe after war was declared.

WORLD WAR I IMPACT ON PENSACOLA

Once war was declared, Naval Aeronautic Station Pensacola experienced unimaginable growth in flight training requirements. Replacements for qualified aviators transferred to combat duty was coordinated by the removal of aviation detachments from ships and the transfer of these pilots to Pensacola and other training Naval air stations soon to be formed.[29] Kite balloons were retained aboard ship as observation platforms and for gunfire

---

[c] The USS JUPITER was later converted into the U.S. Navy's first aircraft carrier, the USS LANGLEY.

spotting. In 1916, Pensacola's training capacity was less than 40 pilots per year but by war's end more than a thousand pilots had been trained here. Seaplanes assigned increased from 39 to 215, officers from 58 to 438, and enlisted personnel from 431 to 5,559.[30]

Construction at the station also exploded as facilities were built to support the increased number of pilots, machine gunners, observers, maintenance men, ground crew and aeroplanes assigned to the station for training. More than 200 seaplanes, dirigibles, and balloons were hangared at the station.[31] Hundreds of additional structures, from hangars to barracks, were built and two outlying airfields, Camp Saufley on Santa Rosa Island for gunnery practice and Camp Bronson at Magnolia Point for aerial bombing practice, were established. [32]

NAVAL AVIATION IN WORLD WAR I

While the performance of U.S. Army aviators in World War I has been well publicized, the achievements of Naval aviators are often overlooked. From a single Naval Aeronautical Station at Pensacola, Naval aviation expanded to more than 31 air stations in Europe and Canada, 14 air stations at home, and 17,500 personnel sent overseas. Between June 1917 and December 1918, only 18 months, U.S. Navy and U.S. Marine Corps aviators flew over 22,000 antisubmarine warfare patrol and bombing flights with 39 attacks on suspected German submarines and more than 100 tons of bombs dropped.

John Ficklen's painting of Lieutenant David S. Ingalls in action with his Sopwith Camel. The original painting hangs in the National Museum of Naval Aviation in Pensacola. (Author's collection)

Lieutenant David Ingalls, flying a Sopwith Camel with British Squadron 213, became the only U.S. Navy ace of World War I when he downed his fifth German aeroplane, a 2-seat Rumpler on September 24, 1918. As Assistant Secretary of the Navy for Aeronautics from 1929-1932 he was a frequent visitor to Pensacola, often flying his own aeroplane.

FLIGHT TRAINING EXPANDS BEYOND PENSACOLA

The requirement for Naval aviators eventually became so large that Pensacola could not train them all and additional Naval air training stations were built throughout the United States and Canada. In May 1918 Pensacola's mission changed from providing elementary flight training to conducting advanced training such as aerial gunnery, bombing, navigation, and night flying.[33] Student aviators received initial flight training at other air stations before reporting to Pensacola.

Figure 5. Camp Saufley was the U.S. Navy's first outlying airfield. Established from 1917-1918 it was located directly across from NAS Pensacola on Santa Rosa Island. (Courtesy National Museum of Naval Aviation)

U.S. NAVY'S NEED FOR OUTLYING FIELDS

The increased demands of World War I flight training also created the first need for outlying airfields away from the saturated Naval Aeronautic Station complex.

Camp Saufley[d] is the earliest known aviation-related camp at Pensacola and was established in the summer of 1918 on Santa Rosa Island. Located across from the Naval Aeronautical Station on an uninhabited beach donated by the War Department, it was used "to afford a place where aerial gunnery instructions could be carried on without danger to the neighborhood."[34] It originally consisted of tents, but by November 1918 wooden buildings were constructed to house a mess hall, gunnery and instruction school, beach houses, radio station, armory, magazine, and a pier for boat landings.[35] Except for night watchmen, all seaplanes and personnel returned to the air station at night.

A second aviation-related camp, named Camp Bronson,[e] also was established in the summer of 1918 at Magnolia Bluff, located approximately eight miles north of the Naval Aeronautic Station on leased land.[36] Used for aerial bombing practice, several buildings were constructed, including barracks, a camera observer house, garage, mess hall, and a boat pier. A steel observation tower over 64 feet high that had been removed from the Naval Aeronautic Station was rebuilt on the site. A summer resort pavilion on the site also was repaired and remodeled, presumably for office space. A note mentioned that "this is not a permanent acquisition of the Air Station, but will return to the owner of the land when the lease expires June 30, 1919."[37]

Figure 6.   The observation tower at NAS Pensacola. (Courtesy National Museum of Naval Aviation)

[d] Named after Lieutenant Richard C. Saufley, Naval Aviator 14, who was killed on Santa Rosa Island on June 9, 1916 while trying to break his own record for hydroaeroplane endurance.

[e] Named after Lieutenant Clarence K. Bronson, Naval Aviator 15, who was killed on November 8, 1916 during a premature bomb explosion while in flight.

NAS OBSERVATION TOWER

During World War I a remarkable number of barracks, mess halls, workshops, storehouses, and concrete paving projects were completed. One item, in particular, evolved into a landmark. Construction of a skeleton steel framework observation tower located on the southeastern corner of the station began in July 1917.[38] In September 1917, *The Pensacola Journal* reported:

Foundations for the great observation tower have been laid, and the steel has arrived. The tower will be 240 feet high, with an electric elevator running within forty feet of the top. It will be fully equipped with telephonic connections to other parts of the station, searchlights for spotting planes at night, and a complete signaling outfit. The base of the tower will be 45 feet, gradually tapering off at the top. [39]

Completed in March 1918, the tower provided observers a clear view of Pensacola Bay to monitor flight operations. Without radios or any other means of notifying the station of an aeroplane in trouble, an immediate emergency response to an aeroplane mishap most times depended solely on the sharp eyes of lookouts posted in the observation tower.

MAGNOLIA BLUFF TOWER

Considering the size of Pensacola Bay, it is not surprising that the establishment of a tower to cover the northern half of the bay was considered prudent. One individual who took a particular interest in this subject was Dr. Blocker of Magnolia Bluff. Dr. Blocker was trying to interest the aeronautic station in the establishment in his neighborhood of "...a sort of an auxiliary to the mother station of the naval air branch."[40] The U.S. Navy agreed and in July 1918 announced that it planned to build a steel pier, 70-foot galvanized observation tower, quarters, and mess room at Magnolia Bluff.[41] After years of service, the tower was dismantled in 1928 and transferred to the Forestry Service.[42]

BOATS, LAUNCHES & DORIES

The station's transportation department was responsible for rescuing both downed flyers and salvaging their damaged flying machines. They had at their disposal dories and launches for towing aeroplanes with minor trouble and speed boats with medical personnel for quick reaction in case of accidents with injuries. They also owned one of the most unique marine derricks in the U.S. Navy for picking up wrecked aeroplanes. [43]

MARY ANN

Of the many firsts at the Pensacola Naval Aeronautic Station, there is none more original than that of the derrick *Mary Ann*, the first marine derrick in the history of the U.S. Navy designed to be self-propelled. She could rush to the scene of an aeroplane accident under her own power "...to lift founded craft out of the water and return them to the station, or to a battleship, without injuring them."[44] Prior to *Mary Ann's* arrival, overturned aeroplanes were

Figure 7. The derrick *Mary Ann* in action in Pensacola Bay. (Courtesy Pensacola Historical Society)

towed back to the station "...a process which in the past has damaged the machine more than the accident itself."[45]

The *Mary Ann* was constructed in Charleston, South Carolina and towed in August 1916 to Pensacola by the collier ARETHUSA for outfitting.[46] She was described as having "... two pontoons, connected by trusses and decking, on which a derrick with a five-ton capacity and hoisting speed of five feet a minute is installed."[47]

The *Mary Ann* was affectionately named after Miss Mary Ann Carter, the daughter of an officer stationed at Naval Aeronautic Station Pensacola.[48]

SAFETY AND RECOVERY

This interrelated system of observation towers, fast boats and *Mary Ann* took care of any crashes at Pensacola. *The Pensacola Journal* described the safety system in effect at the Naval Aeronautic Station:

We have an excellent system for taking care of crashes in the bay. A steel tower one hundred and eighty feet high has been erected at a point overlooking all parts of our flying courses. Four men, in the charge of a petty officer, are on watch in this tower at all times while planes are in the air, each man constantly scanning a quarter of the horizon. No plane is permitted to get out of range of vision from the tower, so that, when a crash occurs, it is sighted almost before the plane strikes the water. Instantly the crash bell rings in the dispensary and at the dock. Doctors and hospital corpsmen rush to the dock where a forty-knot sea-sled is waiting with engine running. Within a minute after the crash the sea-sled with doctor and attendants on board is tearing across the bay to the scene of the accident, followed by the "Mary Ann," a floating crane, to pick up the wreckage. Boats are stationed at different points in the bay to render assistance. [49]

Figure 8. The Sea Sled. One aviator noted that during their first solo flight student pilots were required to fly a prescribed course around Pensacola Bay so the rescue sled could follow them around the course. Consequently, the sled was nicknamed the "solo coffin" and students noted "you could look down and see this thing following you all around the bay." (Courtesy National Museum of Naval Aviation)

POST WORLD WAR I DRAWDOWN

The demobilization of Naval aviation after the end of World War I was quick in terms of closing air stations, canceling aeroplane contracts, recouping unexpended funds and releasing personnel from active duty. After a wartime high of 14[f] Naval air stations in the continental United States, the Fiscal Year 1920 Naval Appropriations Act would limit heavier-than-air Naval air stations to six: Rockaway, Anacostia, Hampton Roads, Key West, Pensacola, and San Diego.[50] Of these, San Diego and Hampton Roads would prove to be Pensacola's greatest competitors to retain Naval aviation training.

That operations at Naval Air Station Pensacola would be reduced upon the conclusion of World War I was a foregone conclusion, even for the most optimistic of Pensacolians. It was just a question of how deep the cuts would go.

On the bright side, all elementary flight training returned to Pensacola from the other Naval air stations. Unfortunately, the decrease in flight training requirements and assigned personnel reduced appropriations for Pensacola to almost nothing and activity at the Naval Air Station came to a standstill. Newspaper articles declared "Pensacola Remains Foremost Air Station" while tacitly admitting that "Some nominal reduction is to be made in the naval force and activities at the Pensacola Navy Station, in keeping with the peace time curtailment of activities to be effected at all the navy yards and naval stations."[51]

Understandably made oversensitive by the closure of the Pensacola Navy Yard only eight years before, front-page headlines in *The Pensacola Journal* constantly reassured Pensacolians that the Naval Air Station was not going to be abandoned. An admiral inspecting the station in March 1919 announced, "Without fear of contradiction I can say that this will be one of the principal, permanent aviation bases to be maintained by the government."[52] He went on to add that "You have every favorable facility to the maintenance of a naval aviation base."[53]

But this was not enough. No one less than the Secretary of the Navy was required to respond to a persistent rumor that closure was imminent. He stated: "With reference to the report that the work would be discontinued at the Pensacola and Key West navy yards, the secretary of the navy expresses surprise that any one should suppose that the work would be discontinued at these yards, and states that it will go on as usual."[54]

Thrown once more into the breach, civic leaders sought new missions for the Naval Air Station to replace the steady decline in training dollars and to reduce the station's vulnerability to closure.

---

[f] Chatham, Montauk, Rockaway, Cape May, Hampton, Yorktown, Moorehead City, Charleston, Brunswick, Miami, Key West, Pensacola, Galveston, and San Diego.

BIRTH OF THE BARNSTORMERS

At the end of the war, both the U.S. Navy and the U.S. Army declared thousands of their aeroplanes surplus, flooding the market with cheap landplanes and flying boats available to all comers at public auctions. *The Pensacola Journal* noted "Even if the peace congress should decide on universal disarmament, there are still any number of uses to which airplanes can be put in time of peace."[55] The first aviation entrepreneurs and barnstormers were about to appear in Pensacola.

NOTES

[1] Bright Outlook For The Pensacola Yard. (1914, May 5). *The Pensacola Journal.*

[2] Bright Outlook For The Pensacola Yard. (1914, May 5). *The Pensacola Journal.*

[3] Preparing For Enlargement Of Flying School. (1914, November 25). *The Pensacola Journal.*

[4] General Navy Board's Request for $5,000,000.00 For Aeronautic School. (1914, December 12). *The Pensacola Journal.*

[5] Turnbull, A.D. & Lord, C.L. (1949). *History of United States Naval Aviation.* New Haven, CN: Yale University Press, pages 47, 81 and 107.

[6] Pensacola As Air Station Much Favored. (1917, December 10). *The Pensacola Journal.*

[7] Will Manufacture Aeroplane Parts. (1915, August 15). *The Pensacola Journal.*

[8] Many Flying Machines Are Being Built. (1915, December 13). *The Pensacola Journal.*

[9] Will Soon Be Fifty Machines Here. (1916, September 28). *The Pensacola Journal.*

[10] More Hangars Are Needed To House The Airplanes Here. (1917, January 5). *The Pensacola Journal.*

[11] Jahncke, E.L. Station Built for a Navy Yard. *Air Station News, Pensacola, Florida.*

[12] More Hangars Are Needed To House The Airplanes Here. (1917, January 5). *The Pensacola Journal.*

[13] Open Bids For Steel Hangars Navy Yard. (1915, October 19). *The Pensacola Journal.*

[14] Navy Yard Presents a Busy Scene. (1916, September 11). *The Pensacola Journal.*

[15] Moving Warrington Houses Rapidly. (1916, August 26). *The Pensacola Journal.*

[16] Three New Hangars To Be Constructed. (1916, September 28). *The Pensacola Journal.*

[17] Completion Of Dozen Wooden Hangars At Navy Yard. (1917, September 16). *The Pensacola Journal.*

[18] Why Pensacola Wings. (1916, October 31). *The Pensacola Journal.*

[19] Naval Commission Tells Why Pensacola Should Be Developed. (1917, March 16). *The Pensacola Journal.*

[20] Anderson, J.H. (1916, September 11). Navy Yard Presents A Busy Scene. *The Pensacola Journal.*

[21] Anderson, J.H. (1916, September 11). Navy Yard Presents A Busy Scene. *The Pensacola Journal.*

[22] Van Wyen, A.O. (1969). *Naval Aviation in World War I.* Washington, D.C.: U.S. Government Printing Office, page 8.

[23] MacCloskey, M. (1968). *From Gasbags to Spaceships.* New York: Richards Rosen Press, page 87.

[24] Ramsey, D.C. (1940, February). Growth of Bureau of Aeronautics Appropriations. *Aero Digest,* page 74.

[25] Boyne, W.J. (1984). *DeHavilland DH-4.* Washington, DC: Smithsonian Institution Press, page 48.

[26] Boyne, W.J. (1984). *DeHavilland DH-4.* Washington, DC: Smithsonian Institution Press.

[27] Sitz, W.H. (1930) *A History of U.S. Naval Aviation.* Washington, DC: Government Printing Office, page 41.

[28] Shirley, N. C. (2000). *United States Naval Aviation 1910-1918.* Atglen, PA: Schiffer Military History, page. 104.

[29] Reynolds, C.G. (1991). *Admiral John H. Towers.* Annapolis, MD: Naval Institute Press, page 115.

[30] Coletta, P.E. (1985). *United States Navy and Marine Corps Bases, Domestic.* Westport, CN: Greenwood Press, page 470.

[31] U.S. Naval Air Station Will Celebrate Centennial Soon. (1925, November 8). *The Pensacola Journal.*

[32] Coletta, P.E. (1985). *United States Navy and Marine Corps Bases, Domestic.* Westport, CN: Greenwood Press, page 470.

[33] Bennett, F.M. (1919, March 15). *Historical Data, Collection of.* Special Collections, 73-6, Box 2, John C. Pace Library.

[34] Bennett, F.M. (1919, March 15). *Historical Data, Collection of.* Special Collections, 73-6, Box 2, John C. Pace Library.

[35] Bennett, F.M. (1919, March 15). *Historical Data, Collection of.* Special Collections, 73-6, Box 2, John C. Pace Library.

[36] Bennett, F.M. (1919, March 15). *Historical Data, Collection of.* Special Collections, 73-6, Box 2, John C. Pace Library.

[37] Bennett, F.M. (1919, March 15). *Historical Data, Collection of.* Special Collections, 73-6, Box 2, John C. Pace Library.

[38] Bennett, F.M. (1919, March 15). *Historical Data, Collection of.* Special Collections, 73-6, Box 2, John C. Pace Library.

[39] Completion Of Dozen Wooden Hangars At Navy Yard. (1917, September 16). *The Pensacola Journal.*

[40] Air Station Branch To Be Located Here. (1918, June 16). *The Pensacola Journal.*

[41] Building Branch Of The Air Station. (1918, July 22). *The Pensacola Journal.*

[42] Navy Gives Up Tower For Use By Forest Men. (1928, February 21). *The Pensacola Journal.*

[43] Transportation. (1922, December 5). *Air Station News, Pensacola, Florida.*

[44] Marine Derrick For the Recovery Stranded Planes. (1916, August 29). *The Pensacola Journal.*

[45] Derrick for Salvaging Airplanes is Launched. (1916, December 16). *The Pensacola Journal.*

[46] Derrick For Salvaging Airplanes Is Launched. (1916, December 16). *The Pensacola Journal.*

[47] Marine Derrick For the Recovery Stranded Planes. (1916, August 29). *The Pensacola Journal.*

[48] Lazarus, W.C. (1951). *Wings in the Sun.* Orlando, FL: Florida Press, page 53.

[49] Raby, J.J. (1924, December 28). Naval Aviation Training. *The Pensacola Journal.*

[50] Turnbull, A.D. & Lord, C.L. (1949). *History of United States Naval Aviation.* New Haven, CN: Yale University Press, page 171.

[51] Pensacola Remains Foremost Air Station. (1919, August 27). *The Pensacola Journal.*

[52] Permanency of Aviation Here Is Sure. (1919, March 21). *The Pensacola Journal.*

[53] Permanency of Aviation Here Is Sure. (1919, March 21). *The Pensacola Journal.*

[54] Reports That Navy Yard Would Close Are Absolutely Unfounded. (1919, July 8). *The Pensacola Journal.*

[55] After War Uses For Airplanes. (1918, November 21). *The Pensacola Journal.*

CHAPTER 12
FIRST COMMERCIAL AVIATION IN PENSACOLA

Figure 1. Published on December 19, 1909 in *The Pensacola Journal*, this is an "artist's depiction" of what the Wright Flyer would have looked like flying over Plaza Ferdinand had the Wright brothers selected Pensacola as the home for their first flying school. (Courtesy *Pensacola News Journal*)

INQUIRIES

As early as 1909, several companies made overtures to Pensacola regarding the establishment of various aviation enterprises here. Nearly every occurrence was announced in a front-page article in *The Pensacola Journal*, although rarely was additional information published later.

WRIGHT BROTHERS AND PENSACOLA

In the winter of 1909, the Wright brothers were looking for a temporary warm-weather site to locate a flying school. Having established the Wright Company aeroplane factory in November 1909, the Wright brothers were ready to penetrate the civilian market with their first practical Wright Flyer. The way to do this, they decided, was by training their own pilots "…to engage in exhibition flying and earn the prize money various organizations were now offering those who achieved aviation firsts."[1] However, with rival aeroplane makers such as Glenn Curtiss entering the market, they could not afford to wait for warmer weather in Ohio to establish a flying school to train their aviators.[2] Accordingly, Wilbur Wright toured Florida, Georgia and Alabama looking a suitable site. A careful aviator and even shrewder businessman, he not only sought a location with historically minimal winds necessary for the light aircraft of this time, but he also expected financial concessions from the local community.

With its sandy soil, steady sea breezes and vast impenetrable forests of pine, Pensacola was not truly an ideal location for a landing field. There is no known record of Wright visiting Pensacola in 1909, but the photograph in Figure 1 indicates that Pensacola was at least aware that he was touring Florida looking for possible training sites.

Wilbur did visit Montgomery, Alabama. A local plantation owner offered the Wrights free use of his land, businessmen offered to build a hangar and clear all land free of charge, and even the free loan of a new automobile was offered for local transportation. After reviewing average wind speed and rainfall graphs, which were ideal, Wright accepted Montgomery's offers and wired Dayton to prepare their flying machines for shipment via train.

Too late to matter, shortly afterwards Wright "...received thirty telegrams from cities throughout the South offering him 'every expense and monetary bonus' if he selected their location."[3] Montgomery had achieved historic success by using overwhelming community support and the lure of free services to entice the Wrights to establish their flying school there. This lesson was not lost on Pensacola businessmen when they would try to draw other aviation activities here.

On March 26, 1910 Orville Wright launched from a field three miles west of Montgomery,[a] the first flight of the first flying school in the United States.

SECOND WRIGHT INQUIRY

In November 1915, Pensacola was seriously considered as the site of another Wright brothers flying school. The Wright Machine Company received a contract for the training of more than 1,000 Canadians in support of World War I preparations.[4] A representative of the company arrived in Pensacola "...in search of a field of at least 240 acres, level and cleared of all trees and stumps, and if such a site is available, free of cost."[5] If Pensacola could provide the site within two weeks and pay for the clearing of a landing field, it would be leased for six months. In return, the Wrights also dangled the carrot that if "...all conditions are satisfactory it is possible the Wright Company may purchase a training site as well as locate a branch factory." [6]

A suitable site of about 200 acres was found near Goulding, located several miles north of the City of Pensacola, and offered to the Wrights. The Wright Company representative "...seemed fairly pleased with it. The entire tract

---

[a] The present site of Maxwell Air Force Base.

is practically covered at present with a small growth of black jack, but if it is acceptable to the Wright people this will be cleared off and the field put in shape by citizens for use as an aviation school."[7] As before, several cities were in competition and eventually the training was located at Dayton, Ohio during the summer and Augusta, Georgia in the winter.[8] A Wright representative later revealed that Pensacola was rejected due to unacceptable wind velocity.[9]

MOISANT INTERNATIONAL AVIATORS

John B. Moisant, an American sometimes called the "King of the Aviators," was one of aviation's true pioneers.[10] Among his many aeronautical accomplishments, in August 1910 he became the third pilot to fly over the English Channel. He was killed in an aeroplane accident at New Orleans on December 31, 1910.[b]. Shortly before his death, Moisant had formed an exhibition team called the Moisant International Aviators, which would continue in operation for the next several years. In August 1912 the Moisant International Aviators contacted Pensacola regarding the possibility of establishing an aviation school, possibly as early as November 1, 1912.[11] There were plans to bring five flying machines to Pensacola and the company requested 2,000 square feet of smooth, level ground for their airfield and two buildings 40x50 feet for hangars.[12] Nothing more can be found concerning Pensacola, but a Moisant school eventually opened on Long Island, New York.

INTERNATIONAL AIRSHIP COMPANY

In November 1916 the International Airship Company of Illinois sent the Pensacola Chamber of Commerce a telegram asking "What advantage would Pensacola offer for aeroplane factory and flying school?"[13]

The Chamber responded:

Equable climate, level back country, excellent transportation facilities, normal temperature 68 degrees; normal wind velocity ten miles per hour, from southwest. Average 255 clear days per year. Skilled labor from U.S. Aviation Station and site on Pensacola bay, suitable for land and water flying, at reasonable cost.[14]

The president of the company responded in a letter by stating that:

I read your letter and circular with a great deal of interest and it is my personal belief that   Pensacola would be as satisfactory a place as any city which has yet made us a bid, from a natural environment standpoint.

Several cities in the South, including New Orleans, have intimated in their correspondence with this company that they would provide a location for this company in case we decided to establish schools and plants in their cities.

---

[b] At one time New Orleans International Airport was named Moisant Field in his honor.

I am not saying this as a feint to bring about a proposition from your organization, but rather to show that Pensacola will have some competition. We are not necessarily looking for a gift, but all other things being equal it would be only natural that the company would consider more favorably any propositions that might carry with them some demonstration on the part of the city making the offer which would show a spirit of cooperation.[15]

Shades of the demands of the Wright brothers! No record could be found of Pensacola's reply but it appears evident that in early commercial aviation if a city wanted to attract an aviation business then they must be prepared to pay for it.

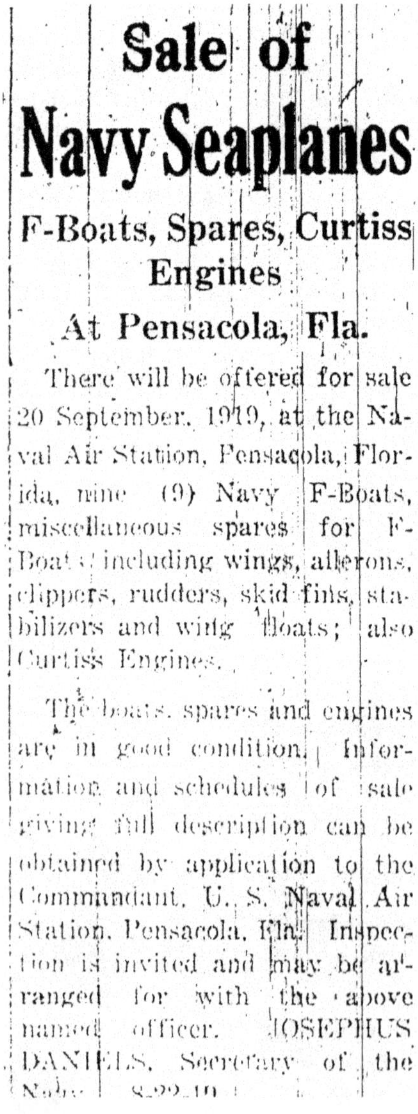

Sale of Navy Seaplanes

F-Boats, Spares, Curtiss Engines

At Pensacola, Fla.

There will be offered for sale 20 September, 1919, at the Naval Air Station, Pensacola, Florida, nine (9) Navy F-Boats, miscellaneous spares for F-Boats including wings, ailerons, clippers, rudders, skid fins, stabilizers and wing floats; also Curtiss Engines.

The boats, spares and engines are in good condition. Information and schedules of sale giving full description can be obtained by application to the Commandant, U. S. Naval Air Station, Pensacola, Fla. Inspection is invited and may be arranged for with the above named officer. JOSEPHUS DANIELS, Secretary of the Navy. 8-22-19.

Figure 2. 1919 advertisement in *The Pensacola Journal* for the surplus sale of U.S. Navy seaplanes. (Courtesy *Pensacola News Journal*)

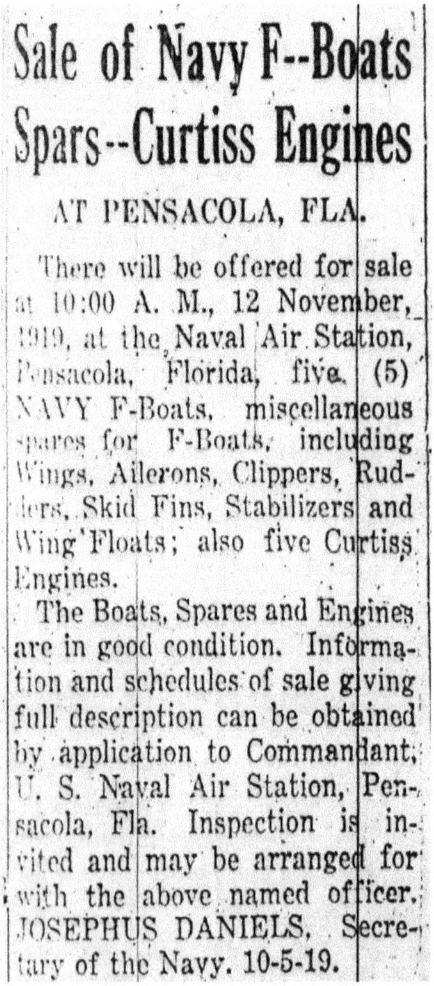

Sale of Navy F--Boats Spars--Curtiss Engines

AT PENSACOLA, FLA.

There will be offered for sale at 10:00 A. M., 12 November, 1919, at the Naval Air Station, Pensacola, Florida, five (5) NAVY F-Boats, miscellaneous spares for F-Boats, including Wings, Ailerons, Clippers, Rudders, Skid Fins, Stabilizers and Wing Floats; also five Curtiss Engines.

The Boats, Spares and Engines are in good condition. Information and schedules of sale giving full description can be obtained by application to Commandant, U. S. Naval Air Station, Pensacola, Fla. Inspection is invited and may be arranged for with the above named officer. JOSEPHUS DANIELS, Secretary of the Navy. 10-5-19.

Figure 3. Another ad for the 1919 sale of U.S. Navy flying boats and parts at Pensacola. (Courtesy *Pensacola News Journal*)

POST-WORLD WAR I AVIATION SURPLUS

The aviation buildup in support of World War I saw the United States with several thousand aeroplanes on hand or on order at war's end. The services cancelled some orders, destroyed some aeroplanes and then simply put the rest up for sale with ads across the country. Newspapers opined on "What will America do with her immense airplane fleet and her army aviators after the war?" and "Will the machines be junked and the aviators be turned loose to seek other purposes?" [16] They hoped not and proposed several uses for the surplus aeroplanes, including flying the mail, coast guard patrol and dispatch work.

While some of these aeroplanes were old, tired training machines, others were brand new, with brand new engines and parts to accompany them. All were sold at a fraction of their acquisition price. Imagine simply strolling onto a Naval Air Station at the end of the Cold War, plunking down less than a third of an F-14's value, and then simply flying it away. The U.S. Army and the U.S. Navy were doing this with their surplus landplanes and flying boats.

FIRST COMMERCIAL SEAPLANE OPERATION IN PENSACOLA

If the U.S. Navy in Pensacola saw fit to train thousands of young men in the art of flying, then there apparently was no reason they could not be outfitted with their own flying boats when they left the service. In Pensacola, a dashing Naval aviator by the name of Lieutenant J. Albert Whitted seized this opportunity to establish Pensacola's first commercial seaplane operation. A native of St. Petersburg, Whitted received his initial flight training from Tony Jannus, an aviation pioneer who established the United States' first scheduled airline service from St. Petersburg to Tampa in 1914.

Figure 4. A former Naval aviator and flight instructor at NAS Pensacola, J. Albert Whitted established the first commercial aviation venture in Pensacola. (Courtesy of author)

Whitted joined the U.S. Navy in March 1917, qualified as Naval Aviator Number 179 in December 1917, and remained at NAS Pensacola as a flight instructor.[17] Eventually rising to the position of chief flight instructor, he made the first flight from Pensacola to New Orleans in a seaplane in 1918 and was in charge of a long distance flight to Cuba in March 1919.[18] By the time of his release from active duty on June 30, 1919, he had accumulated more than 760 hours of flight time.

AIRCRAFT for Business Use

THE adaptability of aircraft to practical business purposes has become an established fact. Business men are coming to look on this useful, creative method of travel as essential in the operation of a successful up-to-date organization. The reason for this acceptance is the proven DEPENDABILITY of the more recent types. The NAVY, through the stress and strain of War, has learned valuable lessons which incorporated in the construction of seaplanes and flying boats, have produced marvels of DEPENDABILITY. They are offering several hundred of these latest model seaplanes, many of them still in the original packing cases, at phenomenally low prices. These seaplanes and flying boats having passed rigid NAVY inspection, are probably the most dependable type of aircraft in the world. They are not toys nor experimental jobs, but are similar to the types used by the NAVY today. They have low fuel cost per passenger mile, long flying range, and are able to withstand the waves and weather. These 'planes are —

ALL NEW — NEVER HAVE BEEN FLOWN

[small print describing seaplanes and flying boats, models, engines, balloons, and ordering terms]

ORDERS WILL BE ACCEPTED FROM THIS ANNOUNCEMENT.

Bureau of Supplies and Accounts
NAVY DEPARTMENT
Washington, D. C.

Figure 5. While this surplus sale did not indicate whether any of the aeroplanes were available in Pensacola, the assortment was certainly impressive enough: HSs, H-16s, F-5Ls, AeroMareine 39Bs, Curtiss Pushers, Boeing Tractor Seaplanes, and a captive balloon. (Courtesy *Pensacola News Journal*)

In October 1919 he purchased a Curtiss flying boat from the U. S. Navy and christened it the *Blue Bird*.[19] He then started his commercial seaplane business in Pensacola and operated it from Runyan's slip at the rear of the Pensacola Yacht Club building.[20] *The Pensacola Journal* was excited to report that "...Pensacolians who for years have had to watch naval flyers will now have their first chance to enjoy the thrills of flying."[21] Whitted maintained his ties with St. Petersburg, contracting for exhibitions, passenger flying, and flight training there during the winter.[22]

A Lieutenant Wheeler, another ex-Naval aviator who purchased a Curtiss flying boat, joined Whitted in his business.[23] The two lieutenants made front-page news during a fundraiser:

Lieutenants Whitted and Wheeler will give a certain percentage of their receipts derived from their flying boat, to the Red Cross during the remainder of the Red Cross Roll Call, according to an announcement made last night by the aviators and Red Cross officials.

The aviators will charge $15 for a twenty-five mile ride around the bay and city and a percentage of this fee will be given to the Red Cross. Tickets may be procured from any of the Red Cross workers at any of the stands. The aviators said last night that they would also take up passengers at night and that their plane will be equipped with a red cross to distinguish it from the naval planes, and that this cross will be lighted at night. They are also making arrangements to shoot Very's lights during their night flights.

Many citizens are availing themselves of the opportunity to take a hop in an airplane and each day Lieuts. Whitted and Wheeler are working overtime in order to give everyone a ride.[24]

The next day's paper reported the results:

Pensacola was highly excited last night when the city was bombarded from an airship, which sailed over the city sending off red, white and blue fireworks, the pyrotechnics providing nothing more disastrous than Red Cross propaganda. A searchlight from the "Bluebird" showed the colors of the red cross, which were displayed in two streamers from the airship.[25]

Figure 6. A Gulf Coast Air Line flying boat at its departure gate. (Courtesy Pensacola Historical Society)

Other barnstorming aviators would appear in Pensacola, including Eddie Nirmaier, an Italian officer who flew with the Royal Italian flying corps, and Eddie Scarlett, another former Naval aviator who also established a commercial flying camp in Pensacola.[26]

WOOD AERIAL CORPORATION

On March 9, 1921 a representative of the Wood Aerial Transportation Company approached the chamber of commerce regarding "The inauguration of an air line from Pensacola to Mobile, from Mobile to New Orleans, and from Atlanta to Mobile, Pensacola and New Orleans."[27] He indicated the company would use flying boats along the coastline and charge passengers 100% more than a railroad ticket, an incentive due to the speed of the aeroplanes. Nothing further was heard of this request.

GULF COAST AIR LINE

On March 26, 1923 *The Pensacola Journal* reported the inauguration of airline service between Mobile and Pensacola by Gulf Coast Air Line.[28] Managed by J. Arthur Newcomb, a former Naval aviator, and based out of Mobile, they owned two flying boats and planned to make Pensacola their home base once more planes were purchased.[c] The airline's vision was to "Build up a regular transportation line that will embrace all of the gulf cities, from New Orleans to Tampa. At first the Mobile-Pensacola link will be developed, probably taking in Camp Walton, Valparaiso and Panama City as the next link."[29]

---

[c] For all the news this airline would generate in *The Pensacola Journal*, a search of *The Mobile Register* did not reveal a single article.

The operation was apparently not year round as in November 1923 *The Pensacola Journal* reported:

The Gulf Course (sic) Airline will resume operations to and from Pensacola next summer, giving the city direct communication by air with New Orleans and Mobile, according to announcement yesterday by J. Arthur Newcomb, manager.

Three new six-seated flying boats have been purchased by the company for the purpose. Mr. Newcomb said the business which our company did last summer between Pensacola and Mobile justifies the belief that such a service next summer will be well patronized.

'We are receiving inquiries occasionally concerning trips to Tampa, and even as far as Havana, all of which is an indication that people are beginning to realize the advantages of air travel. If the opportunity presents we shall make a flight from New Orleans to Pensacola some time in the future and we will prbably (sic) extend the flight on to Tampa.'[30]

History has not been kind to Gulf Coast Air Line. One article stated that "This company is the second in all the United States to put commercial flying on a regular schedule. It is the first to install an all-year service."[31] Unfortunately, not a single aviation history book acknowledges this achievement.

Nonetheless in January 1924 the "first non-stop commercial flight" between New Orleans and Pensacola took place when a six-passenger flying boat of Gulf Coast Air Line arrived in Pensacola after a two and a half-hour flight.[32] Allen S. Hackett was identified as president of Gulf Coast Air Line with "...a number of prominent New Orleans business men..." as stockholders.[33] Gulf Coast Air Line also "...operates a mail and passenger service between New Orleans and Pilottown, which is 90 miles south of the Crescent City and a famous duck shooting place."[34]

In what was perhaps the first delivery of newspapers by plane, on July 4, 1923 *The Pensacola Journal* chartered Gulf Coast Air Line to deliver an extra newspaper detailing results of the Dempsey-Gibbons fight. The paper reported:

With large bundles of the fight extras in the cockpit, one of the Gulf Coast Air Line's seaplanes hopped off at Bayou Chico shortly after the final result and was speeding toward Camp Walton and Valparaiso. Ray Fredrichsen, of the circulation department, made the hop to distribute The Journals; Albert Whitted piloted the plane, Harry Thompson was mechanician.[35]

It is interesting to note that Whitted is now flying for Gulf Coast Air Line, as is Harry Thompson, the former Army aviator.

TRIANGLE AIRWAYS OF CHICAGO

The arrival in December 1921 of an HS-2L commercial seaplane belonging to Triangle Airways of Chicago is typical of the many visits by commercial airlines to Pensacola. Hosted at the Naval Air Station, aboard was the airline's President, J. A. Colvin, who announced that:

...the trip was a 'feeler' as to the prospects of commercial aviation, and that his company was contemplating the establishment of an airline between New Orleans and Chicago, provided it was found to be practical. We believe that we could establish the fact that air travel is comfortable and safe. We feel that our trip south has established that fact for we did not suffer the slightest accident en route.[36]

The seaplane returned to New Orleans with several passengers and no other record of this company can be found in Pensacola.

SOUTH FLORIDA AVIATION PASSENGER SERVICE COMPANY

In September 1925, an announcement was made from Key West of the "Establishment of a passenger service air line between Key West and several other Florida cities..."[37] At least 25 seaplanes were to be used and the cities to be served included Key West, Miami, Jacksonville, Tampa, Pensacola, West Palm Beach, Punta Gorda and Fort Myers.[38] Nothing further is recorded regarding this company.

FLORIDA AIRWAYS CORPORATION

In September 1925, Florida Airways Corporation announced it planned to establish commercial aviation service in Florida.[39] The airline purchased four Ford-Stout metal transports and a Curtiss Lark plane for use from Atlanta to Jacksonville, Miami, Fort Myers, and Tampa. A stop in Pensacola was planned when these aeroplanes were shipped to Florida for service but accidents along the way prevented this.[40]

Florida Airways Corporation became the first airline to be issued a State of Florida airplane transportation license and the first Florida airline to win a federal airmail route when they were awarded Contract Air Mail 10 on February 11, 1926. While a planned expansion plan would have provided service to Pensacola, the airline went out of business by the end of the year, a victim of insufficient mail and passenger revenues.[41]

METAL AIRCRAFT EXPRESS COMPANY

G. Douglas Dardrop, secretary of the Metal Express Company, announced that his company was "...arranging for a daily airplane and seaplane service between Jacksonville and Havana in November...via Palm Beach, Miami and Key West."[42] The company said it owned 10 all metal German Junker planes, with more under construction in Germany." He arrived in Pensacola in September 1925 and met with Mr. T.C. Imeson, Pensacola City Commissioner, to inspect "...possible locations for a landing field." [43] The article did not clarify whether the company's intention was to include Pensacola on the route, or build a plant here. In either regard, nothing was heard further.

Figure 7. The Savoia Marchetti S-55 flown by the American Aeronautical Corporation to Pensacola in October 1929. (Courtesy *Pensacola News Journal*)

AMERICAN AERONAUTICAL CORPORATION

In October 1929 four airmen from the American Aeronautical Corporation landed on Pensacola Bay in a 16-passenger Savoia Marchetti seaplane to be hosted by the Naval Air Station.[44] Based in Port Washington, Long Island, New York, "the airmen are touring both the inland and coastal waterways of the United States studying the possibilities of seaplane air lines throughout the nation."[45] Nothing further was found on this company.

INTERNATIONAL AIRWAYS CORPORATION

The first hint that civilian aeroplane manufacturing might occur in Pensacola appeared in a brief article in *The Pensacola Journal* in July 1929:

CITY MAY HAVE AIRPLANE PLANT

Cleveland Men May Lease Part Of Shipbuilding Concern

Airplanes manufactured in Pensacola may soon be a reality.

Rumors that a concern might make planes here were confirmed last night by Paul P. Steward, president of Pensacola Shipbuilding Co.
Two airplane men of Cleveland, Ohio have had the matter under consideration for the past six months, he said. Their names are Weiger and Eads (sic). They will be back in Pensacola in the next few days. Their object is to turn out airplanes at the shipbuilding plant, which has been semi-active since the war.[46]

In October 1929 Charles J. Weger, vice president of International Airways Corporation, and his aeronautical engineer, Walter F. Eade, announced that their company had selected Pensacola to manufacture the "Wittemann Follow-plane," a huge three winged, four engined, large capacity aeroplane.[47]

Figure 8. Artist's depiction of flying boat to be built in Pensacola by the International Airways Corporation. (Courtesy *Pensacola News Journal*)

They would use the existing facilities of the Pensacola Shipbuilding Company to construct the first plane. They said "...when quantity production is reached early next spring, it will then be decided whether or not the shipbuilding plant will be enlarged to meet the requirements, or an entirely new site acquired and new factory constructed."[48]

Weger added that in addition to the construction of the flying boat, his company "...plans the establishment of coastwise lines on the Gulf coast from Fort Myers to Corpus Christi, a line from New Orleans up the river to St. Louis, and on the Atlantic seaboard from Miami to Norfolk."[49] This company also had plans for lines to Europe, Mexico, and Central and South America with Pensacola as a base.

Within a year, the Pensacola Metal Aircraft Corporation, as it was now known, raised over $250,000, including the sale to local citizens of $10,000 in notes, to finance construction of one of the flying boats.[50] What became of this company is one of Pensacola's great aviation mysteries of the early 1930's.

Figure 9.   Barely visible on the roof of this building is "Pensacola Metal Aircraft Corp." The company planned to build huge civilian flying boats in Pensacola. (Courtesy West Florida Historic Preservation, Inc.)

PUSH FOR COMMERCIAL SEAPLANES

In December 1928, Admiral J.J. Raby in an address broadcasting over local radio station WCOA (*Wonderful City Of Advantages*) observed that "Pensacola has the finest sheet of water that can be found anywhere for the training of seaplane students."[51]   He urged them to "...take advantage of this fact to promote the manufacture here of seaplanes, and the establishment of commercial seaplane routes as part of the general trend toward transportation through the air."[52] An editorial published the next day echoed the admiral's remarks and exhorted "Now it is up to the people of Pensacola to work for commercial aviation with that same devotion and business acumen that naval officers devote to advancement of the naval air station."[53]

However, the age of the seaplanes was ending.   The future flew in landplanes.

PENSACOLA'S FIRST SCHEDULED AIRLINE

WISE GULF TRAVELERS **FLY** THE GULF-WISE WAY

ATLANTIC & GULF COAST AIRLINE, INC.

For Reservations, Call Atlantic & Gulf Coast Airline—Telephone 4851
—RATED A-1 BY THE UNITED STATES GOVERNMENT—

On August 11, 1935 *The Pensacola Journal* stated that "Pensacola's designation as the 'Annapolis of the Air' and its inability to get a commercial air line forms a paradox which is a poor advertisement for the city. The situation should be remedied at the earliest possible moment..." This request was realized in December 1936 when officials of the Atlantic and Gulf Coast Airline sought an exclusive franchise to operate the only airline service from Pensacola's new Municipal Field currently under construction.

The president of the airline was H. G. Strachan, Jr., the son of H.G. Strachan, Sr., prominent owner of the Strachan Shipping Company. Initial plans called for service between Savannah, Georgia and Mobile, Alabama with enroute stops at Jacksonville, Tallahasse, and Pensacola. The airline flew tri-motored Stinsons with a capacity of eight passengers and 400 pounds of freight.   Arrival times in Jacksonville were coordinated so that passengers could connect with Eastern Airline's northern and southern flights.

In February 1937 10 test flights along the proposed route were conducted by the Department of Commerce to ensure that all pilots and ground crews were properly trained and that all aeroplanes, equipment and radios were in proper working order.

On March 5, 1937 Pensacola's first scheduled airline service began with the arrival and quick departure of an airliner on a westbound flight. The aeroplane spent the night in Mobile and returned the next day. Initial service was three round trips per week and daily service began less than a month later. While nothing further is recorded in *The Pensacola Journal*, aviation history books indicate that the airline went out of business in July 1937, most likely due to lack of an airmail contract and low passenger volume. It was not until 1938 that another airline came to Pensacola to provide scheduled service.

NOTES

[1] Tribble, P. (1997). The Wright Connection. *Alabama Heritage. 45, 6-15.*

[2] Tribble, P. (1997). The Wright Connection. *Alabama Heritage. 45, 6-15.*

[3] Tribble, P. (1997). The Wright Connection. *Alabama Heritage. 45, 6-15.*

[4] Would Train 1,000 Men As Aviators Here. (1915, November 27). *The Pensacola Journal.*

[5] Would Train 1,000 Men As Aviators Here. (1915, November 27). *The Pensacola Journal.*

[6] Would Train 1,000 Men As Aviators Here. (1915, November 27). *The Pensacola Journal.*

[7] Local Field Suitable For Aero School. (1915, November 28). *The Pensacola Journal.*

[8] Molson, K.M. (1963) Canadians at the Wright School. *American Aviation Historical Society, 8(1),* pages 40-41.

[9] Foley, B. (n.d.). *Philip Boyer in Jacksonville, Florida 1915.* Retrieved from http://memers.tripod.com/ralphcooper0/eboyer.htm.

[10] Borden, N.E. (1963). The Remarkable John B. Moisant. *American Aviation Historical Society Journal, 8(1),* pages 3-12.

[11] Aviation Co. May Establish School Here. (1912, August 17). *The Pensacola Journal.*

[12] Aviation Co. May Establish School Here. (1912, August 17). *The Pensacola Journal.*

[13] Aero Factory May Relocate In Pensacola. (1916, November 25). *The Pensacola Journal.*

[14] Aero Factory May Relocate In Pensacola. (1916, November 25). *The Pensacola Journal.*

[15] Airship Company May Investigate Advantages Here. (1916, November 30). *The Pensacola Journal.*

[16] After War Uses For Airplanes. (1918, November 21). *The Pensacola Journal.*

[17] Arthur, R.W. (1967). *Contact!* Washington, DC: Naval Aviator Register.

[18] Pensacola Has First Chance to Enjoy "Hop". (1919, October 22). *The Pensacola Journal.*

[19] Pensacola Has First Chance to Enjoy "Hop". (1919, October 22). *The Pensacola Journal.*

[20] Pensacola Has First Chance to Enjoy "Hop". (1919, October 22). *The Pensacola Journal.*

[21] Pensacola Has First Chance to Enjoy "Hop". (1919, October 22). *The Pensacola Journal.*

[22] Lieut. Whitted Lands Safely. (1920, May 8). *The Pensacola Journal.*

[23] Joy Flying Is To Be Resumed. (1919, November 4). *The Pensacola Journal.*

[24] Aviators Will Divide Revenue With Red Cross (1919, November 7). *The Pensacola Journal.*

[25] Air Ship Flies For Red Cross Roll Call Fund. (1919, November 8). *The Pensacola Journal.*

[26] Pensacola Is On Flying Map. (1920, May 9). *The Pensacola Journal.*

[27] Stone Presents Flying Plans. (1921, March 9). *The Pensacola Journal.*

[28] Record Trip By New Plane Line. (1923, March 26). *The Pensacola Journal.*

[29] Record Trip By New Plane Line. (1923, March 26). *The Pensacola Journal.*

[30] Air Line Is To Operate To And From This City. (1923, November 11). *The Pensacola Journal.*

[31] Air Line Is To Operate To And From This City. (1923, November 11). *The Pensacola Journal.*

[32] Resume Seaplane Service Between Gulf Coast Towns. (1924, January 17). *The Pensacola Journal.*

[33] Resume Seaplane Service Between Gulf Coast Towns. (1924, January 17). *The Pensacola Journal.*

[34] Resume Seaplane Service Between Gulf Coast Towns. (1924, January 17). *The Pensacola Journal.*

[35] Journal Runs Extra on Shelby Scrap; Quick Delivery By Seaplane And Auto. (1923, July 5). *The Pensacola Journal.*

[36] Commercial Plane Is Repaired Here. (1921, December 14). *The Pensacola Journal.*

[37] Million Dollar Syndicate Plans State Air Line With Pensacola Among Stations. (1925, September 19). *The Pensacola Journal.*

[38] Million Dollar Syndicate Plans State Air Line With Pensacola Among Stations. (1925, September 19). *The Pensacola Journal.*

[39] Florida Air Line To Begin This Winter. (1925, September 22). *The Pensacola Journal.*

[40] Airline Vanguards To Come Here. (1925, December 21). *The Pensacola Journal.*

[41] Pensacola To Get Air Line Says Report. (1926, February 10). *The Pensacola Journal.*

[42] Florida Air Line To Begin This Winter. (1925, September 22). *The Pensacola Journal.*

[43] Florida Air Line To Begin This Winter. (1925, September 22). *The Pensacola Journal.*

[44] Airmen Arrive To Talk Over Air Line Idea. (1929, October 9). *The Pensacola Journal.*

[45] Airmen Arrive To Talk Over Air Line Idea. (1929, October 9). *The Pensacola Journal.*

[46] City May Have Airplane Plant. (1929, July 19). *The Pensacola Journal.*

[47] Huge Seaplane To Be Built In Pensacola. (1929, October 27). *The Pensacola Journal.*

[48] Huge Seaplane To Be Built In Pensacola. (1929, October 27). *The Pensacola Journal.*

[49] Huge Seaplane To Be Built In Pensacola. (1929, October 27). *The Pensacola Journal.*

[50] Committee to Push Financing of Air Project. (1930, July 14). *The Pensacola Journal.*

[51] Admiral Looks For Expansion of Air Station. (1928, December 20). *The Pensacola Journal.*

[52] Admiral Looks For Expansion of Air Station. (1928, December 20). *The Pensacola Journal.*

[53] An Aviation Center. (1928, December 21). *The Pensacola Journal.*

CHAPTER 13
KIWANIS FIELD: PENSACOLA'S FIRST AIRPORT

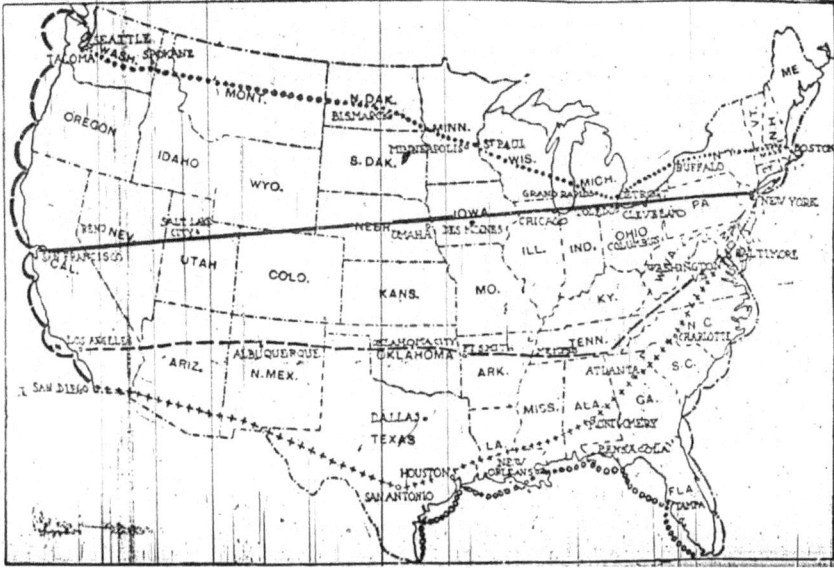

Figure 1. This 1917 proposed airway map shows Pensacola on the "Gulf Airway" which would run from Key West to Texas. The accompanying article emphasized that cities along the route would do well to establish a landing field available for military flyers and the post office. (Courtesy *Pensacola News Journal*)

KIWANIS FIELD

The initiative to establish the first landing field in Pensacola's history originated from a most unlikely source: civilians seeking to support the U.S. Army. The first public mention of a landing field was two sentences in a May 9, 1920 article in *The Pensacola Journal*:

Another aviation stunt which promises to develop into something tangible is a landing field for army aircraft. The Kiwanis Club has endorsed a resolution and it is understood several possible locations have been studied with a view to obtaining a site.[1]

A few days later, another front-page article stated that J. H. Cross had appeared before the Rotarians and asked that they assist the city in selecting a landing field.[2] It was apparently brought to Cross' attention that the U.S. Army was seeking a field for maneuvers near Pensacola and he asked that the various organizations of the city work together toward achieving this goal.

In 1920, all of the U.S. Navy's aviation activities in Pensacola were focused on seaplanes and lighter-than-air operations and there were no U.S. Army aeroplanes stationed at Fort Barrancas. An occasional landplane may have visited Pensacola but they were easily accommodated at the Naval Air

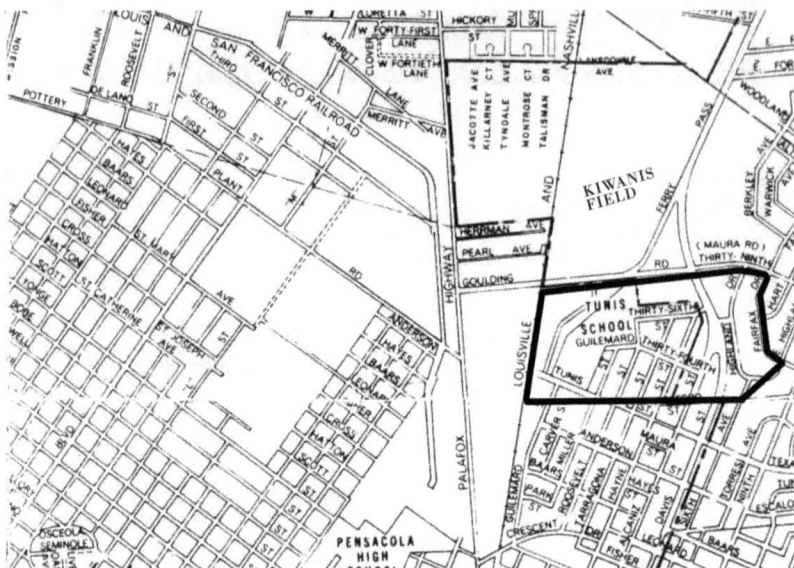

Figure 2. Approximate location of Kiwanis Field is indicated. Pottery Plant Road is now Fairfield Drive and Ferry Pass Road is now Davis Highway. (Author's collection)

Station's balloon field or Fort Barrancas' parade deck. The Kiwanis Club wisely recognized that these fields were controlled by the military and civilians were permitted to land there only as a professional courtesy, a privilege easily suspended. The construction of Pensacola's first civilian landing field would open the door to more than U.S. Army landplane operations. It would provide unrestricted access for commercial operations, including the possibility of an airmail route to the city and placement of Pensacola on the "aviation map."

Financing, selection, construction and opening of the landing field unfolded at a remarkable pace. On May 28, 1920 *The Pensacola Journal* reported:

KIWANIS CLUB FINANCES AIRPLANE LANDING FIELD

APPROPRIATES FUNDS AND AGREES TO SECURE AND PAY FOR CLEAR-ING GROUND

The Kiwanis Club at its weekly meeting yesterday voted to finance an airplane landing field for Pensacola and agreed to take over an option on such a field south of the fertilizer factory at Goulding. The action was taken on motion of Leslie Partridge following a brief but spirited address by Harry Thompson[a] on what the possession of a landing field means to the city.

Mr. Thompson and Mr. Partridge both told of the opportunities ahead for cities which have landing fields available when the aerial routes are mapped out and Mr. Thompson told of the presence in the city of a representative of the Fox News, Inc. a national motion picture news organization.

---

[a] The same Harry W. Thompson, formerly of the U.S. Army Air Service, who stunted over Pensacola during the World War I fund drives.

According to Mr. Thompson, a representative of the film company was in the city yesterday with a proposition to bring at least three flyers to Pensacola within a fortnight if a landing field could be found. These flyers are to take pictures of the city for the Fox people and the pictures will be shown all over the United States. Still pictures will be made by A. Knox McIntyre of Reynald's.

Among the flyers to come here are Capt. Eddie Stinson and William Brock, both of who Thompson knew when he was in the service. Mrs. Katherine Stinson, sister of Eddie, also will come here.

Mr. Thompson made arrangements for the field and for leasing it for one year under an option to buy yesterday morning. He then presented the matter to the club which quickly decided to act, inasmuch as it had been pushing the airplane landing field proposition for some time. The matter was referred to Mr. Thompson by Secretary Price of the chamber of commerce.[3]

On June 1, 1920 construction began on the landing field site located at "Goulding almost due east of the fertilizer factory and about southeast of the turn of the Ferry Pass road after it crosses the railroad track."[4] This was the same site proposed to the Wright brothers in November 1915. Work consisted of mostly of pulling stumps, filling in holes, and rolling the landing field.[5] The field was to be marked by placing four flagpoles in each corner and a newspaper article noted that:

A giant 'P' to indicate Pensacola will be chalked in the center of the field and a wind-direction indicator will be placed at one corner of the field. These marks are in accordance with Aero Club rules and are official. The Aero Club will be informed of the fact that Pensacola has a flying field and the various aviation magazines will make mention of it. The third assistant postmaster general, who has charge of the aerial mail, will also be notified and requested to put Pensacola on the aerial mail route when the southern branch is organized.[6]

On June 25, 1920 the field was formally dedicated as Kiwanis Field.[7] With the C-7 dirigible from the Naval Air Station hovering overhead, several speakers spoke briefly to the assembled picnickers. Colonel F.G. Mauldin of the Coast Defense at Fort Barrancas drew the most attention with his comments that the "War department was anxious to obtain a landing field in Pensacola so that the big guns at the fort could be spotted by airplane."[8] Cross, acknowledged as the primary mover in establishing the field, spoke of "...the need of an airplane field for Army and commercial land craft."[9]

A few weeks later *The Pensacola Journal* exulted:

Pensacola is strategically situated to become a great aviation center. Her harbor is ideally adapted to seaplanes. Kiwanis landing field provides accommodation for land planes. We are ready.[10]

All that was needed to complete the picture was some landplanes, either military or civilian.

USS MASSACHUSETTS

On October 14, 1920 the U.S. Army Air Service announced that they would use Kiwanis Field for the sinking of the obsolete battleship USS MASSACHUSETTS (BB-2).[b] Their airplanes would "...spot the shots from the railway batteries and from the mortars when they are firing..."[11] However, on November 13, 1920 the Army aviators' participation was cancelled and Pensacola lost her opportunity to "...demonstrate to the public the necessity of having a municipal landing field."[12]

This minor setback detracted nothing from the citizens of Pensacola's visionary embrace of the aeroplane's potential, particularly that of the commercial landplane. However, events were transpiring behind the scenes that made the future of flight training in Pensacola bleak.

LANDPLANES FOR THE U.S. NAVY

As late as 1919 the U.S. Navy was on record as stating that Naval aviation did not need landing fields:

Landing fields are not necessary for the naval aviation section but they are for the army. Any place where a beach is seen affords a haven for the pilot of the hydroplane, while the army pilots have to select more suitable fields.[13]

The U.S. Navy's intention was to build an aviation fleet of super-seaplanes and dirigibles to support its fleet and coastal patrol missions. However, all that would change when senior admirals were persuaded to convert an old collier into a revolutionary new "floating aerodrome."

---

[b] The USS MASSACHUSETTS (BB-2) was eventually scuttled in January 1921 in 26 feet of water about a mile and a half southwest of the Pensacola Pass. It is presently designated an underwater archaeological preserve.

NOTES

[1] Pensacola Is On Flying Map. (1920, May 9). *The Pensacola Journal.*

[2] Landing Field Is Again Urged. (1920, May 19). *The Pensacola Journal.*

[3] Kiwanis Club Finances Airplane Landing Field. (1920, May 28). *The Pensacola Journal.*

[4] Construction Work Started On Kiwanis Landing Field. (1920, June 1). *The Pensacola Journal.*

[5] Construction Work Started On Kiwanis Landing Field. (1920, June 1). *The Pensacola Journal.*

[6] Construction Work Started On Kiwanis Landing Field. (1920, June 1). *The Pensacola Journal.*

[7] Kiwanis Field Opening Held. (1920, June 26). *The Pensacola Journal.*

[8] Kiwanis Field Opening Held. (1920, June 26). *The Pensacola Journal.*

[9] Kiwanis Field Opening Held. (1920, June 26). *The Pensacola Journal.*

[10] Journal Delivery Is Made By Plane. (1920,11 July). *The Pensacola Journal Journal.*

[11] Army Flyers Plan to Use Kiwanis Field (1920, October 14). *The Pensacola Journal.*

[12] Postpone Arrival of Air Squadron. (1920, November 13). *The Pensacola Journal.*

[13] Air Station Has Fertile Field To Try. (1919, February 6). *The Pensacola Journal.*

CHAPTER 14
PENSACOLA AND THE USS LANGLEY (CV-1)

Figure 1. USS LANGLEY (CV-1), the "Old Covered Wagon." A former commanding officer described her as "slow and in many ways a lubberly ship, still she trained a generation of Navy pilots to find a flight deck – a chip of home in the wide blue sea." The USS LANGLEY was a frequent visitor to Pensacola. (Courtesy U.S. Navy)

SEAPLANES IN ASCENDANCY

Although the world's first shipboard landing was made by the U.S. Navy and the first three aeroplanes purchased by the U.S. Navy were landplanes, Naval aviation remained committed to the almost exclusive procurement and operation of seaplanes for years. In a world with few landing fields, the benefits of operating seaplanes were numerous. All the major Naval bases were located on the water, water provided an infinite number of places to land, and the construction and maintenance of expensive landing fields were avoided. Further, water was thought to be softer to crash upon than hard ground.

In 1915 the Secretary of the Navy noted:

It is not in the purview of the navy to utilize air craft except in connection with the fleet or for strictly naval purposes. It is because the navy intends to restrict its aircraft to over-sea service or as eyes for the fleet that it has not asked for as large an ap-

Figure 2. Civilian pilot Eugene Ely lands aboard the USS PENNSYLVANIA (ACR-4) on January 18, 1911 in the world's first shipboard landing. The next landing aboard a U.S. Navy ship did not occur until 1922. (Courtesy U.S. Navy)

propriation as would have been required if it had in view as some erroneously suppose, the use of these new scouts of the air on land as well as on water. Land utilization is not a function of the navy. The navy is an institution afloat and must leave land flying to other departments and agencies.[1]

Even in commercial aviation, flying boats ruled the skies as Pan American Airways' majestic Flying Clippers and Imperial Airways' magnificent Short 'C' flying boats pioneered transoceanic routes throughout the world.

In this golden era of seaplanes, few could foresee that by 1945 they would be abandoned, their place taken by the faster, longer-range landplanes who had access to an abundance of large, paved landing fields built during World War II. For the U.S. Navy, however, it was the introduction of the aircraft carrier that paved the way for the large-scale introduction of landplanes into the U.S. Navy. For Pensacola's civic leaders, it presented a formidable challenge to ensure that Naval aviation remained in their city.

U.S. NAVY LANDPLANE TRAINING

Some early Naval and Marine aviators did learn to fly landplanes, it was just that the U.S. Navy did not train them. Instead, the U.S. Army or civilian aeroplane manufacturers provided the training. As early as August 1915, the U.S. Army agreed to train U.S. Navy and U.S. Marine Corps pilots in landplanes at its Signal Corps Aviation School located at Rockwell Field on North Island at San Diego, California.[2]

During World War I Naval and Marine aviators were shipped overseas to train and operate with British, French and Italian landplane squadrons, which they did with great success. Yet, a post-World War I accounting of all U.S. Navy aeroplanes on November 1, 1919 revealed that of 1,982 total airframes, only 144 were landplanes.[3] The U.S. Navy and U.S. Marine Corps were

Figure 3. The U.S. Marine Corps took the lead in landplane flying within Naval aviation. These are DH-4s assigned to the Northern Bombing Group based in the Calais-Dunkirk area during World War I. Although an agreement had been reached that the U.S. Army would fly landplanes and the U.S. Navy seaplanes during the war, attacks on German submarine bases, considered a Naval mission, were better executed by landplanes flown by Marine and Naval aviators. (Courtesy U.S. Navy)

content to maintain a small cadre of landplanes and receive flight training for these aeroplanes from various U.S. Army Air Service schools such as those at Carlstrom Field, Florida, March Field, California, and Langley Field, Virginia.[4]

However, the British were experimenting with an entirely new method of operating landplanes that would revolutionize warfare at sea.

BRITISH "DECK FLYING"

In this age of the powerful American nuclear supercarrier, few people realize that the British were the true pioneers of aircraft carrier development. During World War I, Great Britain converted several ships into small seaplane carriers.[5] On December 25, 1914 three of these ships launched seven seaplanes on an attempted air raid on the German Zeppelin bases at Cuxhaven. While not entirely successful, one aviation historian noted that "if any single action gave birth to the concept of aircraft carrier operations, this raid would qualify."[6]

Figure 4. The first British shipboard landing was made by Squadron Commander E.H. Dunning aboard HMS FURIOUS on August 2, 1917. With no provisions for an arrested landing, his aeroplane was equipped with underwing handles for deck personnel to grasp and literally pull the aeroplane down. (Courtesy of author)

Recognizing the military advantages of the more maneuverable landplane over the lumbering seaplane, Great Britain next experimented with launching landplanes at sea, outfitting a packet ship with a 64-foot long launching platform. The first successful take off was made on November 3, 1915 and the aeroplane landed alongside the ship for hoisting aboard.

One final challenge remained. On August 2, 1917 a Sopwith Pup landed aboard the flight deck of HMS FURIOUS, the first ship to demonstrate the true capabilities of an aircraft carrier.[a]

This successful operation of landplanes at sea did not escape the attention of Naval aviators but few in the Naval service recognized the potential of carrier aviation. Nearly four years passed before the U.S. Navy was ready to commission their own first carrier.

U.S. NAVY SHIP PLANE UNITS

In the post-World War I era, the U.S. Navy did experiment with launching landplanes from platforms built over a ship's gun turret. The U.S. Navy's first ship plane unit was formed on the East Coast in February 1919, with two

---

[a] The world's first aircraft carrier with a full flight deck did not appear until the HMS ARGUS, a converted Italian passenger liner, was commissioned in 1918. The first aircraft carrier designed from the keel up was also British: the HMS HERMES, commissioned in 1923.

Figure 5. Annotation states "First flight of airplane off an American battleship by Lt. Comdr McDonnell of U.S.S. Texas on March 10, 1919." (Courtesy of author)

landplanes embarked aboard the USS TEXAS.[7] On March 9, 1919 Lieutenant Commander Edward O. McDonnell[b] launched in a Sopwith Camel from a platform built over the USS TEXAS' number two turret to make the first recorded platform take-off.[8]

One Naval aviator provided an excellent description of the platform launching technique:

Launching from the platform was accomplished by raising the tail of the aircraft to level flight attitude with the tail skid placed in a trough, about four feet long. The plane was restrained by a "release hook" until the throttle was fully advanced and the turret was trained so the wind was right down the center line. Then the pilot released the hook. It was found necessary to have at least 22 knots of relative wind down the turret, and even with this amount of wind, the planes sank after leaving the platform until the wheels barely cleared the water.[9]

With no means of recovering these landplanes back aboard, they were flown ashore, broken down and then transported back to the ships. With the ship's guns masked by the launching platform, it soon became apparent that perhaps the use of a catapult and seaplanes was a more optimal solution. Nonetheless, additional ship plane unit's were established on both coasts and their pilots, trained by the U.S. Army, flew a bewildering hodgepodge of landplanes.

---

[b] Another former Naval aviator, Forrest E. Wysong, claimed to have flown a Sopwith Camel off the USS TEXAS in November 1918, before McDonnell's attempt. However, the U.S. Navy can neither confirm nor deny this event.

NAVAL AVIATION'S FIRST AIR-CRAFT CARRIER

Although the U.S. Navy had demonstrated the first successful take-off and landing of an aeroplane from a ship,[c] there was little support for the development of a dedicated aeroplane carrier. However, the British carrier exploits of World War I convinced the U.S. Navy that they needed to experiment with an aircraft carrier of their own. Utilizing funding authorized on July 11, 1919 the collier USS JUPITER (AC-3) sailed to the Norfolk Naval Shipyard and was decommissioned on March 24, 1920 to begin a two-year conversion program into the U.S. Navy's first aircraft carrier. She was recommissioned on March 20, 1922 as the USS LANGLEY (CV-1) and within a few months she was ready for flight operations. The USS LANGLEY was about to change the U.S. Navy's entire perspective on the value of landplanes, and whether the U.S. Army should continue to provide landplane training to Naval aviators.

Figure 6. The USS LANGLEY pierside at NAS Pensacola. (Courtesy *Pensacola News Journal*)

FLIGHT OPERATIONS ABOARD THE USS LANGLEY

The officer responsible for training the USS LANGLEY's first pilots was Lieutenant Commander Chevalier. He had been tasked by Commander Ken Whiting, prospective executive officer of the USS LANGLEY, with forming a pre-commissioning aviation detail for the USS LANGLEY. Chevalier was to use the turret-flying pilots and mechanics from his ship plane unit, the Naval aviators with the most current landplane experience. In the spring of 1921, this unit off-loaded their aeroplanes at the newly completed, grass-covered landing field at Naval Air Station Hampton Roads, Virginia. [d]

In turn, Chevalier tasked Lieutenant Alfred M. Pride with solving the problem of how to trap aboard an aircraft carrier. Pride had been trained in landplanes by the U.S. Army and as he recalled later:

---

[c] On November 14, 1910 civilian pilot Eugene Ely took-off from a wooden platform built on the bow of the USS BIRMINGHAM (CL-2). On January 18, 1911, he trapped aboard the USS PENNSYLVANIA (ACR-4). Ely was killed in a barnstorming accident at Macon, Georgia on October 19, 1911.

[d] Coincidentally, NAS Hampton Roads' landing field was located directly in front of Commander Theodore Ellyson's, Naval Aviator Number 1, quarters.

Figure 7. The USS WRIGHT (AV-1) was initially capable of both lighter-than air and heavier-than-air aviation support. (Courtesy U.S. Navy)

Chevalier walked up to me one day in the summer of 1921 and told me to stay ashore at Norfolk and begin work. Meanwhile, the Navy Department had built a turntable on the field at Hampton Roads, Virginia, about 100 feet in diameter, flush with the ground. They mounted on some gear that the British were then using, but that gear was not successful, and they ripped it out. In its place, I mounted crosswires, instead of sandbags, that were designed to carry along weights suspended in towers. The arresting experiments didn't take very long, probably not more than six months. I had to hustle because I understood the Langley was going into commissioning the spring of 1922.[10]

The USS LANGLEY'S pilots were soon making their own landings aboard the turntable and also practiced landing on a 60-foot long cloth strip placed on the field that represented the USS LANGLEY's flight deck.[11] Finally, all was ready for actual underway flight operations and on October 26, 1922 Chevalier made the first trap aboard the USS LANGLEY. U.S. Navy aircraft carrier operations had begun.

Figure 8. USS LANGLEY at Pensacola with the two dirigible hangars and station landing field visible in the upper left corner. (Courtesy National Museum of Naval Aviation)

USS WRIGHT (AV-1)

Surprisingly, the first ship totally dedicated to aviation operations in the U.S. Navy wasn't an aircraft carrier at all. Although little known, the USS WRIGHT was a unique aviation ship in her own right, the first tender in the U.S. Navy used exclusively for both lighter-than-air and heavier-than-air operations. Named for Orville Wright, she was commissioned on December 16, 1921 as the AZ-1, both a lighter-than-air capable ship and seaplane tender.[e] Initially equipped with a balloon well from which to launch kite balloons, she was converted to a heavier-than-air aircraft tender in 1926 and redesignated the AV-1. She supported seaplane operations on the eastern seaboard and Caribbean Sea and was a frequent visitor to Pensacola.

On April 19, 1923, the *Air Station News* noted a historic occurrence:

The Wright arrived yesterday after an absence of nine months. The Langley being here also makes this a very notable event as this is the first time that the Langley and the Wright have been together in this harbor. With the combined resources of these two ships any Navy would be completely equipped for battle against any foe, as the Wright in addition to being a tender to seaplanes, is also a Lighter-than-Air carrier. Kite balloons may be used for spotting for the Fleet as well as the planes, while the Langley may use the fast land planes as spotters, taking off and landing on deck of the ship while at sea.[12]

U.S. NAVY SEEKS ITS OWN LANDPLANE TRAINING

Returning to the fall of 1921, a year before the USS LANGLEY conducted her first flight operations, the U.S. Navy reviewed its policy on having the U.S. Army provide Naval aviators with landplane training. In view of the large demand for qualified pilots to fly from the decks of her new aircraft carriers,[f] the U.S. Navy was now ready to assume responsibility for the training of its own landplane pilots.

While landplane training started immediately in Pensacola, it was not a foregone conclusion that this new landplane training would stay in Pensacola. The Secretary of the Navy knew very well that there were no landing fields at Naval Air Station Pensacola and he was not prepared to invest a lot of money in building them. With Naval appropriations getting smaller and smaller, he needed a Naval air station that could accommodate landplane training with a minimum investment of his limited resources. Naval aviation was preparing to leave its cradle for its birthplace.

---

[e] It is interesting to note that while the U.S. Navy named its first aircraft carrier, USS LANGLEY, in honor of Professor Langley, the first totally dedicated aviation ship in the U.S. Navy, USS WRIGHT, was named after one of the Wright brothers.

[f] Both the USS LEXINGTON (CV-2) and USS SARATOGA (CV-3) were authorized for completion as aircraft carriers on July 1, 1922. They would be commissioned on December 14, 1927 and November 16, 1927 respectively.

LANDPLANES FOR THE LANGLEY

Exactly which landplane would be embarked aboard the USS LANGLEY presented somewhat of a problem as there were none specifically designed for carrier operations. Consequently, three different aeroplanes flew the most significant events in USS LANGLEY's history.

The first take off from USS LANGLEY was made on October 17, 1922 when Lieutenant Virgil Griffin launched in a Vought VE-7SF aeroplane. On October 26, 1922 Lieutenant Commander Godfrey Chevalier made the first landing aboard the USS LANGLEY in an Aeromarine 39B aircraft. On November 18, 1922 Commander Kenneth Whiting made the first catapult launch from the USS LANGLEY in a Naval Aircraft Factory PT torpedo seaplane.

NOTES

[1] Many Flying Machines Are Being Built. (1915, December 13). *The Pensacola News Journal.*

[2] Johnson, E.C. (1977). *Marine Corps Aviation: The Early Years 1912-1940 (G.A. Cosma, Ed.).* Washington, DC: History and Museums Division, U.S. Marine Corps, page 9.

[3] Report of the Navy Department of Aircraft On Hand November 1, 1919. (April-June 1957). *American Aviation Historical Society Journal, 2(s),* page 79.

[4] Grossnic, R.A. (1997). *United States Naval Aviation 1910-1995.* Washington, D.C.: US Government Printing Office, page 42.

[5] World Aircraft Carriers List: RN Seaplane Carriers & Tenders. Retrieved from http://www.hazegray.org/navhist/carriers/uk_sea.htm.

[6] MacDonald, S. (1964). *Evolution of Aircraft Carriers.* Washington, DC: CNO, page 12.

[7] Larkins, W.T. (1996). *Battleship and Cruiser Aircraft of the United States Navy.* Atlen, PA: Schiffer Publishing, page 10.

[8] Dickey, F.C. (1965. U.S. Navy Ship Plane Units. *American Aviation Historical Society Journal, 10(2),* page 121.

[9] Dickey, F.C. (1965. U.S. Navy Ship Plane Units. *American Aviation Historical Society Journal, 10(2),* page 124.

[10] Pride, A.M. (1986). Pilots, Man Your Planes. *Proceedings,* pages 28-35.

[11] Tate, J.R. (1998). We Rode the Covered Wagon. In Wooldridge, E.T. (Ed.) *The Golden Age Remembered.* Annapolis, MD: Naval Institute Press, page 48.

[12] Aircraft Carriers Together Here For First Time. (1923, April 20). *Air Station News, Pensacola, Florida.*

CHAPTER 15
OLD CORRY FIELD

Figure 1. One of the few photographs found of Old Corry Field. Pensacola's first landing field hosted military aviators, barnstormers, flying circuses and various aviation celebrities for several decades. In this instance, a press conference has just been completed with the French aviators Coste and Bellonte, who visited Pensacola in their famous aeroplane *Point d'Interrogation (Question Mark)* shortly after their historic east-to-west across the Atlantic Ocean in 1930. (Courtesy West Florida Historic Preservation, Inc.)

OLD CORRY FIELD

Once described as one of the finest landing fields in the south, Old Corry Field has disappeared, a victim to urban development. Other than a street sign in Warrington located several miles from the airfield's actual site, nothing remains to identify Pensacola's first civilian landing field. Yet its development as a U.S. Navy landing field has to be one of the most important civic tasks ever undertaken by the citizens of Pensacola. Blessed by visionary civic leaders and privately counseled by close friends in the U.S. Navy, Pensacolians acted resolutely and with extraordinary vigor in the fall of 1921 to preserve the U.S. Navy's presence in Pensacola by constructing this landing field.

TWO FIELDS WITH THE NAME CORRY

The original Corry Field discussed in this chapter was located north of Pensacola on land leased by the city. Originally known as Kiwanis Field, this field was presented to the U.S. Navy and dedicated as Corry Field on December 7, 1922. By 1927, increased U.S. Navy landplane training requirements and civilian encroachment around the landing field made flight operations dangerous.

In order to ensure that U.S. Navy landplane training remained in Pensacola, another tract of land was purchased for the U.S. Navy north of the Naval Air Station. On July 9, 1927, this field was dedicated as Corry Field as the U.S. Navy decided to retain the same name. To distinguish between the two fields, the original field then became known as Old Corry Field.

Eventually closed, nothing remains today to identify Old Corry Field. The new Corry Field was closed to aviation operations in 1958 and is the present home to the Naval Technical Training Center.

PENSACOLA INHOSPITABLE FOR LANDPLANES

Situated amidst vast forests of virgin pine, it is no surprise that the primary business of Pensacola in the late 19th and early 20th centuries was the export of lumber and naval stores. In fact, in 1900 Pensacola ranked first in the United States in lumber exports, shipping more than 340 million feet of timber.[1] Thousands of acres were stripped, leaving behind nothing but tree stumps with little thought toward reforestation. Accordingly, when the first Naval aviators arrived in Pensacola they found the surrounding countryside covered in either forests, stumps, or soft sand, all in all a very inhospitable area for landplanes. An early aeroplane mechanic at NAS Pensacola remembered:

...all the planes we received on the station were seaplanes, there were no land planes because there was no place in this general area where you could land planes. I want to make this statement about land planes and land planes landing here; I doubt where there was a space 100 feet square where you could set down a land plane, because there were pine stumps all over this area. [2]

Not that early landplanes needed a large, paved landing field. Aeroplanes were built to land on almost any terrain and the lengths of take-off or landing runs were short. While we are used to 12,000-foot runways capable of launching an 800,000 pound 747, these early planes weighed less than a thousand pounds and could be airborne within a few airplane lengths. Landing distances were equally short. Light aeroplane weights, low landing speeds and the high drag of landing on grass fields even precluded the need for brakes. A Fokker Triplane, for example, needed less than 400 feet to land.[3]

In 1913 the U.S. Army defined a "good landing place" as one offering a runway of at least 200 yards in all directions, so that a safe landing might be made no matter in which direction the wind was blowing.[4]

NAS NORTH ISLAND TO CONDUCT
ALL U.S. NAVY LANDPLANE TRAINING

As noted earlier, the closure of the Pensacola Navy Yard in 1911 made Pensacolians exceedingly insecure about the permanence of their Naval air station. Anything that affected the station, from floods to hurricanes to the post-war demobilization, was cause for anxiety. Pensacolians would fret and worry after every incident until reassured by senior U.S. Navy authorities that their fears were unfounded and that the Naval Air Station would remain open. However, in the case of landplane training for the U.S. Navy, there was ample reason for concern. Secretary of the Navy Edwin Denby already had made a decision not to use Pensacola for landplane training.

The reasons were financial in nature. While rich in tactical aircraft carrier development initiatives, the U.S. Navy was cash poor due to the steep decline in naval appropriations after World War I. Rather than spend money on the construction of new landing fields at Pensacola the Secretary of the Navy preferred to use the existing facilities of Naval Air Station North Island near San Diego, California.

NAVAL AIR STATION NORTH ISLAND

Self-described as the "Birthplace of Naval Aviation," an honor well-deserved, Naval Air Station North Island, California traces its aviation heritage to the establishment of a Curtiss Aviation School on its uninhabited shores in January 1911. Lieutenant Theodore G. Ellyson, Naval Aviator Number 1, was trained to fly here by Glenn Curtiss. Naval aviation left North Island in May 1912 and did not return until 1917 when World War I training requirements dictated the need for a permanent naval air station on the west coast. The reporting of her first commanding officer, Lieutenant Earl W. Spencer,[a] on September 25, 1917 to establish a seaplane base is considered the birthday of Naval Air Station North Island.[5]

Meanwhile, U.S. Army aviators had been on North Island since November 2, 1912 and by December 1913 had consolidated all of their aviation activities here, designating their landing field as Rockwell Field. Expansion of U.S. Army flight training at Rockwell Field during World War I was enormous, with nearly 141 landplanes based at the field in 1918.

While the U.S. Army conducted landplane operations, the U.S. Navy focused solely on seaplane training. Yet as early as 1919 Lieutenant Spencer envisioned training Naval aviators in landplanes here. He pointed out to the Chief of Naval Operations that "...the North Island area was of the best sand and clay soil, firmly packed, covered with short grass and perfectly level" and that "...North Island had ample space for construction of landplane hangars."[6]

Lieutenant Spencer's vision became a reality when the Pacific Fleet was re-established in June 1919 with San Diego as the principal Naval base. To support this fleet, Air Detachment, Pacific Fleet was commissioned in October 1919 and three divisions were formed: landplane, ship-plane and seaplane.[7] Seaplanes were already at the station and lighter-than-air activity soon began with balloons and airships. In October 1919, landing

Figure 2. In preparation for operations aboard the USS LANGLEY, a "Ship-plane Practice Landing Platform" was built at NAS North Island in 1921 for practice carrier landings. (Courtesy of author)

fields were prepared at the Naval Air Station and landplanes were borrowed from Rockwell Field to begin training. Prospective landplane pilots first completed the normal course of instruction in seaplanes at Pensacola before

---

[a] In 1916, Lieutenant Spencer married Wallis Warfield, whom he met while stationed in Pensacola. After their divorce in 1927 and a brief marriage to Ernest Simpson, she became involved with King Edward VIII, who later abdicated the British throne to marry her.

reporting to North Island for instruction in landplanes.[8] NAS North Island's splendid landplane training infrastructure could now easily absorb all of the U.S. Navy's landplane training requirements with a minimum of expense.

PENSACOLA ACTS

With perfect hindsight, we know now that if Pensacola had lost landplane training to North Island the subsequent decline in the importance of the seaplane to the U.S. Navy would have resulted in the eventual abandonment of all flight training at Pensacola. However, a joint committee of the Kiwanis Club, Rotary Club, Retail Merchants Association and Chamber of Commerce members, using the sage advice of Naval officers with close ties to Pensacola,[b] met to discuss the issue. Together they devised a means of permanently retaining landplane training in Pensacola that was simple in its design, flawless in its logic, and nearly effortless in its ease of execution.

In one of the most brilliant business decisions ever made in Pensacola's history, they agreed to secure leases on a tract of land suitable for a landing field and then offer it free of charge to the U.S. Navy. As had been done years ago for the U.S. Army and Kiwanis Field, how could anyone possibly turn down a free landing field? Once the U.S. Navy saw that suitable outlying landing fields were readily available to disperse landplane training, they would immediately appreciate the economies of scale offered by collocated seaplane and landplane training in Pensacola.

While the estimated cost to the city would be $1,500, nearly $15,152 in 2002 dollars, these business leaders knew that the possible return on this investment over the years would be measured in the millions of dollars.[9]

LANDING FIELD SITE SELECTED

Two possible landing field sites presented themselves.[10] However, the site at Goulding located "…about five miles north of the city near the fertilizer plant" was the more suitable of the two.[c] It was certainly the most familiar. It was the same location originally offered to the Wright brothers and the current home of Kiwanis Field.

As Kiwanis Field was deemed too small for the U.S. Navy's use, adjacent property would have to be leased to expand the total size of the proposed landing field. A note in one of the committee member's personal papers gives a good idea of the cost involved:

---

[b] Captain Henry Mustin, Admiral Moffett's Assistant Bureau Chief and first commander of the Naval Aviation detachment upon its arrival in Pensacola in January 1914 traveled to Pensacola several months before Admiral Moffett's visit. It was probably he who suggested a strategy to Pensacolians to convince Admiral Moffett to locate landplane training here.

[c] The other site was described as being west of the city. While a plat is known to exist of this site, it cannot be located but it is presumed that this is the current site of Corry Field. This field also would be used as an emergency landing field.

I have acquired already leases on all of the land except Miss Lola's, under conditions that the Chamber of Commerce pay an annual rental to the amount of the taxes now on the land, in figures this amounts to be about $1.50 an acre, and the leases already procured contain the option to purchase this land, at a valuation of $150.00 per acre. These are the prices which are agreeable to all of the other property owners, and I believe are equitable and right. [11]

Once the leases were all in order, all that remained was a pro forma offer of the landing field to the Secretary of the Navy to formalize acceptance of this unprecedented offer.

U.S. NAVY DECLINES THE LANDING FIELD

On September 21, 1921 Pensacola sent a letter to the Secretary of the Navy offering the landing field and inquiring "...if this landing field is turned over to your department for the purpose mentioned what assurance can be given us that it will be used by the navy department." [12] The response was not long in coming. On October 11, 1921 the Secretary of the Navy replied:

Gentlemen:

The Navy Department is appreciative of the offer of the Chamber of Commerce of Pensacola. Fla., as made in its letter of September 21, 1921. However, such land plane training as may be necessary probably will be conducted by the Navy at the Naval Air Station, San Diego, Cal. At that place it is expected that there will be adequate facilities. Hence, to maintain land planes at Pensacola would constitute duplication out of keeping with the present need for economy.

Very truly yours,

Edwin Denby [13]

Pensacola's leaders may have been stunned but they were not about to be denied.

ADMIRAL MOFFETT VISITS

The Secretary of the Navy's letter arrived only a few days before Rear Admiral William A. Moffett, the Chief of the Bureau of Aeronautics, made an official visit to Pensacola as part of a joint committee established to select landing fields for the U.S. Navy and U.S. Army. Admiral Moffett would prove to be an influential figure and good friend in the development of both Naval and commercial aviation at Pensacola. But right now he was the single most important person

Figure 3. Rear Admiral Moffett, Chief of the Bureau of Aeronautics, flanked by his Naval aviator sons George on the left and William on the right. (Courtesy of author)

who might influence the Secretary of the Navy to reconsider his decision on landplane training basing.

Upon his arrival on October 24, 1921 Admiral Moffett visited the two landing field sites and afterwards stated that he "...had found both very good, but one possibly more sandy than was desirable."[14] However, the Admiral's next comment, as quoted in *The Pensacola Journal*, was the most important:

...if the government could find it possible to expend the money necessary for the fields, he should be glad to recommend Pensacola for this purpose, but that he could not say at this time what would be done, as the appropriation for action had been so closely cut, the aviation bureau having 25 per cent less this year than last, for expenditures.[15]

The message was clear: the U.S. Navy did not have the money available to purchase land for landing fields. In view of the fact that their free offer of a landing field had just been turned down, the civic leaders of Pensacola must have been scratching their heads in wonder.

However, Admiral Moffett did not leave them without hope. He was further quoted as saying "owing to the fact that most of the flight training of the navy is done here.... he considered this the most important naval air station in the country."[16] *The Pensacola Journal* noted that "...the fact that Admiral Moffett looks upon it [Naval Air Station Pensacola] as the most important station in the country, may perhaps be accepted as augury for future development."[17] Pensacola had a very powerful friend on their side, augmented by the influence of their Congressmen and Senators. However, could the decision of the Secretary of the Navy be overturned solely on the recommendation of Admiral Moffett?

SECRETARY OF THE NAVY RECONSIDERS

On December 12, 1921 the Pensacola Chamber of Commerce received a very different letter from the Secretary of the Navy that stated:

Gentlemen:

Referring to my letter of October 11, 1921 in reply to your letter of September 21st, I have to inform you that the situation in regard to Naval land plane training has been changed since my reply.

I therefore ask if you will be kind enough to renew the offer in your letter of September 21st, and further to inform me if Naval land planes may be permitted to use your field for landing and taking off without expense to the Government other than costs that might be incurred by damage to property in accidents.

If it is found practicable for the Navy to use your field, it should be made ready for use shortly after July 1st next. The extent of our contemplated operations there cannot be stated definitely at present, but is expected that the field would be used on all routine flying days through the year.

Very truly yours,

Edwin Denby[18]

Pensacola could not respond quickly enough. *The Pensacola Journal* ecstatically reported on December 13, 1921 that the government had "...accepted the landing fields offered the United States Navy by a joint committee of the Chamber of Commerce, the Kiwanis and Rotary clubs on the occasion of the visit of Admiral Moffett to Pensacola several months ago."[19] Leases were immediately arranged and the next step was preparation of the field for flight operations, and the "official acceptance" of the landing fields by the government, guaranteeing their use to establish a landplane flying school.

ADMIRAL MOFFETT RETURNS

In May 1922, Admiral Moffett returned to Pensacola to take the five-week Naval Aviation Observer Course.[d20] On the eve of his departure, he addressed a luncheon of the Kiwanis Club and impressed upon his listeners:

...the fact that without aviation the Pensacola navy yard would be of little moment, and reminded them of the strenuous efforts which have been made to have the naval air station removed from here to other locations, among which he mentioned Hampton Roads and Birmingham.[21]

Admiral Moffett also added that "...there was no doubt but that the Naval Air Station would have been moved from Pensacola had it not been for the two landing fields, which Pensacola had offered the government."[22]

The city fathers had done their work well. At the same luncheon, Harry W. Thompson noted that the annual payroll at the station was more than $1 million a year.[23]

FIELD PREPARATIONS

In tribute to the efforts of the Kiwanis Club whose initiative brought about the field, Admiral Moffett decreed that the U.S. Navy would retain the name Kiwanis Field.[24] What condition the field should be in for landplane flying was clearly defined by the commandant of Naval Air Station Pensacola in a letter to the Chamber of Commerce:

1. In replying to a verbal request from representatives of the Chamber of Commerce, you are informed that the field should be cleared of all trees, scrub oak and under-brush. Wherever grass appears it should not be longer than grass after being cut by a regular grass cutter.

2. The field should be cleared to an extent of approximately one mile by one-half mile. This will be of sufficient size for present operations.

3. All abrupt changes in contour should be removed and holes filled. Wagon ruts and small wood roads across the field should be scraped over and filled in.

4. A gently rolling surface, a gentle rise, or a gentle slope is not objectionable, for example, the surface of the present Kiwanis field, where level, is satisfactory.[25]

---

[4] The observer course was designed to familiarize non-aviator senior Naval Officers with Naval aviation so that they could fill designated aviator billets.

The Pensacola Journal described the work being done to the field:

With a clearing fully one mile in length and more than a half mile wide, the Kiwanis landing field near Goulding is practically ready for use by the United States government in connection with the land plane training recently added to the course in aeronautics at the Pensacola Naval Air Station.

For several weeks a force of about fifteen men with several teams and a tractor have been clearing the scrub oaks off the tract and leveling the ground. These are employed by the city. The board of county commissioners furnished a crew of men for about two weeks and they cleared a considerable plot of ground but they abandoned the work some time ago, leaving the city to finish the job.

An old railroad cut across the field is now being filled by putting the scoops, drawn by pairs of mules, into action. A part of the crew is busy completing the clearing of the oaks, burning trash and rolling the field. Preparation of the field is being done under the direction of Phil Sanchez, street superintendent.

Framework for collapsible plane hangars has already been placed on the field by the forces at the Naval Air Station and these will be erected during the next week, it is understood. The hangars will be on the order of those used by the expeditionary forces during the recent war, but as the land flying activities here are increased it is expected that permanent hangars will be erected.

No attempt to surface the field with clay is being made. It is said that the instructors do not care to have this field made too easy for the student flyers. A hard-surfaced field has been in use at the station for several months and it is from this field that those who have completed the regular seaplane course and launch into the land plane flying get their first practice. As they advance in experience they will attempt (sic) *to the Kiwanis field which will be more difficult to train from. 'Plenty of space is what we want' is what one of the officers from the station said in this connection. And when one drives along Palafox road and looks out across the clearing is* (sic) *appears that their wants have been fulfilled.*

The field, however, Mr. Sanchez said yesterday, will be put into the best condition possible without applying a surface of clay.

While the field is being put into shape especially for the government it is probable that as the use of the airplane becomes more common and they invade the south in the commercial field that Kiwanis field can be used by them. This day, however, is some time off.[26]

FINDING OLD CORRY FIELD

References in older books state that Old Corry Field was located near Washington High School. However, that school was relocated to a new facility on Airport Boulevard.

The old high school has become the J.E. Hall Educational Services Center on 30 East Texar Drive. Another excellent reference point is the Brown Barge Middle School or Uncle Bob's Self Storage.

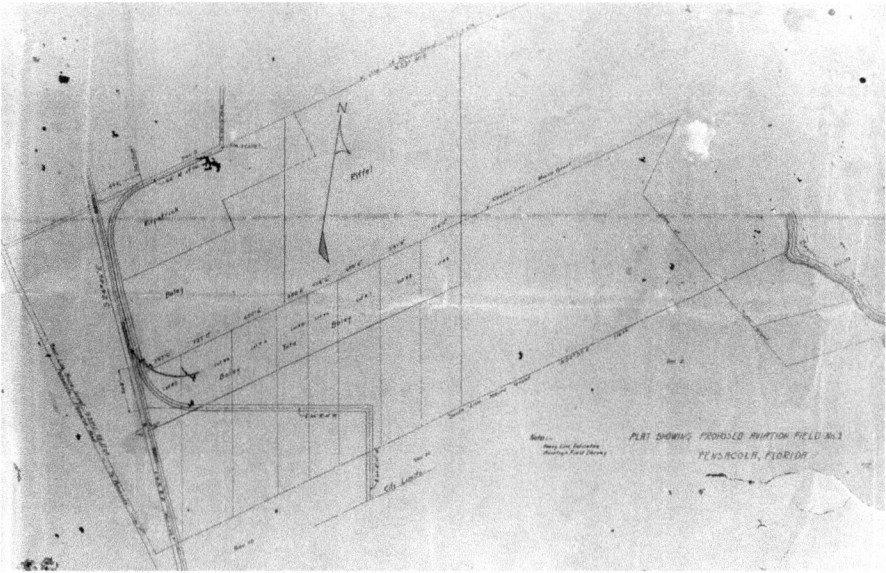

Figure 4. Only known plat showing the exact position of Old Corry Field, Pensacola's first landing field located north of the city in Goulding. (Courtesy University of West Florida Special Collections Library)

CORRY FIELD DEDICATED

After a year of preparation, the field was ready to be officially dedicated by the U.S. Navy. The Kiwanis Club recommended that the field be named Corry Field, in honor of Lieutenant Commander William M. Corry, Naval Aviator No. 23, who was the first Floridian to be designated as a Naval aviator.[e] The U.S. Navy agreed.

On December 7, 1922, Corry Field was dedicated:

CORRY FIELD IS DEDICATED WITH IMPRESSIVE CEREMONY

Governor, Naval Officers and Prominent Citizens Take Part in Exercises

Airplanes circling high in the air dropping flowers, as symbols of the navy's respect and admiration, smoke from a salute of 21 guns hovering over a crowd estimated at 5,000 and marked by addresses emphasizing unselfish sacrifice, service and patriotism, Corry Field was impressively dedicated in memory of Lieutenant Commander William M. Corry, Jr., yesterday afternoon.

Varying from the clear weather of the past two weeks, the day was one to lend solemnity to the occasion, to attune the spirit of the assemblage to the spirit of reverence, love and memory. It was raining at 1 o'clock but when the motorcade arrived at Corry Field rain had ceased to fall leaving the skies leaden.

---

[e] Born in Quincy, Florida, Lieutenant Commander Corry died on October 7, 1920 from injuries received while trying to rescue a fellow pilot from a flaming crash both were in. He was subsequently awarded the Congressional Medal of Honor for his heroics.

Figure 5. In this 1935 aerial view of Pensacola Old Corry Field's parallelogram shape is clearly visible to the north of the city. Pensacola Municipal Airport, now known as Pensacola Regional Airport, is not as defined but is located approximately east of this field across the bayou. (Courtesy West Florida Historical Preservation, Inc.)

With detachments of sailors and marines from the naval air station, the air station band, a detachment of soldiers from Fort Barrancas leading, the parade formed in front of the San Carlos hotel at 2 o'clock. The honor guests of the day, representative of the city, county, air station, civic organizations and others followed in automobiles.

400 Cars Take Part

The parade headed down Palafox street to Zarragossa, east on Zarragossa to Jefferson, north to Government, west to Palafox again, north to Wright, and east to Tarragona where the sailors, soldiers, marines and many others boarded a special train for Goulding. Hundreds of cars joined in the motorcade, approximately 400 taking part.

At the field the speaker's stand was banked on either side with six land planes. Across the field, lined to perfect symmetry, were five others. Later these rose from the spacious aeronautical training ground, flew gracefully above, thrilling those present with innumerable stunts.[27]

Speeches were given by the governor of Florida and the commandant of the Naval Air Station, who praised the "patriotism and public spirit of the citizens who have made this splendid tract of land available to the navy."[28] The commandant also noted that a layman "…may be excused if he expresses surprise at the navy's need of land-type airplanes" and that if a prophet had told him that one day he would command a flying field he would have had "…grave misgivings as to the prophet's sanity."[29]

OLD CORRY FIELD IMPROVEMENTS

While the dedication ceremony may have suggested the beginning of flight operations, the U.S. Navy already was using the field and an aviation support detail from the Naval Air Station was living and working in tents on site to support flight operations. Official acceptance by the U.S. Navy occurred in September 1923 when a party of Naval officers and Pensacolians inspected the entire field and pronounced it satisfactory. This meant that "the present temporary quarters of the men will be replaced by substantial barracks for the mechanical force, fuel stations, and water supply stations will be enlarged and everything put in first rate condition for intensive training."[30]

The U.S. Navy also planned to seed the sandy field in order to "…aid the growth of grass and reduce the cutting up of the field by airplanes."[31] Fencing was planned for the site and it was noted was that:

Fertilization of two hundred forty acres of the new tract has been completed by the Public Works Department, and if the cows and the lovers can be discouraged from parking thereon, we should have an excellent airdrome before many moons.[32]

Trespassing was evidently a major issue. A later newspaper article noted that:

…The navy now exercises complete jurisdiction over Old Corry Field, situated near Goulding, and trespassers on the property will be reported to the proper authorities. There has been considerable driving of autos and trucks across the field, but this will

Figure 6. The only reminder of Old Corry Field is a road named in its memory in Warrington. This road is not located anywhere near the original site. (Courtesy of author)

not be tolerated in the future. Trucks driving over the field tend to tear up the soil and make ruts which are dangerous to the flyers in attempting to make landings.[33]

Apparently, progress was slow on the promised construction projects. A reporter visiting Old Corry Field in May 1923 noted that a hangar constructed shortly after the dedication was "...condemned as unsafe for the storage of planes..."[34] He also reported that a permanent detail assigned there "...live in tents, have their own mess, a man to cook for them, and nothing to bother them except mosquitoes and the maidens from the vicinity who pay them assiduous court. All hands have nets over their bunks so that danger is lessened."[35]

Other pilots reported that the:

...open sandy field had 'no facilities whatever' other than a single tent. In the winter months, students stood around an open fire keeping their hands warm while waiting for their next flight. If one landed in a 'too heavy' piece of sod, the plane would not roll and it would go up on its nose. Fortunately, the wooden propellers could be rapidly replaced.[36]

1924 DESCRIPTION OF OLD CORRY FIELD

In 1924 a news journalist visiting Corry Field described it for his readers:[37]

The immensity of Corry Field is not realized when one gives it a casual glance. Its average length is 4,215 feet while the average width is 2,770 feet. It is oddly shaped, being more of a parallelogram than a rectangle. Its area is approximately 265 acres. It has excellent approaches from all sides and is of such size that pilots do not need to land close to the edges. The surface is sandy, covered with grass, and gives resiliency to the landing planes. In point of size, desirability from an aviation standpoint, drainage, facilities, etc., it ranks favorably with the leading aviation fields of the country. It is unexcelled in the entire south and the army aviation field at San Antonio, Texas (Kelly Field) is probably the only one in the south that ranks with it.

Contrary to general belief, Corry Field is not closed to aeroplanes other than navy planes. It has a municipal and commercial standing in the aircraft year book publish by the Aeronautical Chamber of Commerce of America. The rating given it is very good. Since the field has been established approximately a dozen civilian planes have made use of it.[38]

In March 1925 reporters noted several improvements at the field:

Preparations are being made to house and feed the large number of men necessary for the handling and upkeep of the planes. A new barracks and mess hall has been erected on the field. A heating system has been installed that will heat both water and oil, making it easier to start the machines in cold weather. Another improvement at the

---

Pensacola.—Old Corry Field, municipal, rating ——. Two miles N. of Pensacola. Altitude, 75 feet. Rectangular, 3,200 by 2,205 feet, sod, level, natural drainage. Smoke stacks to NW. and W. Facilities for servicing aircraft, day only.

Figure 7. 1934 U.S. Bureau of Commerce description of Old Corry Field. (Courtesy Chris Kennedy)

field is a machine gun test stand where the guns will be tested before and after they are fired from the planes while in flight.

OLD CORRY FIELD OPERATIONS

Practically all of the U.S. Navy's landplane training was conducted at Old Corry Field. Flight students first completed seaplane training at the Naval Air Station, including ground and aerial gunnery, elementary bombing, and navigational flights before moving to Old Corry Field for landplane training.[39] However, neither Naval authorities nor the City of Pensacola intended that Old Corry Field was reserved for the exclusive use of Naval aviators. The commandant of the Naval Air Station pointed out that:

While Corry Field was designed primarily by the navy as an inducement for land plane training here, it will also be available for use by commercial planes. It will be necessary, however, for commercial aviators desiring to land there to first get permission from the naval authorities at the naval air station.[40]

For the next several decades, student Naval aviators, barnstormers, flying circuses, cross country travelers, visiting businessmen and dignitaries shared Old Corry Field.

One example of non-military use was the storage of a plane purchased by U.S. Navy pilot AMM1c L.G. Muller for his own use. He:

...purchased a JN4D plane equipped with an OX5 Curtiss motor at Montgomery, Alabama, during the later part of April, and ferried it to Old Corry Field, where he kept it until his discharge on 17 June. Muller had a flying proposition offered him at Boulder, Colorado and it was his idea to keep the machine here, get it in the best possible condition and then fly in it to Colorado where he would install a Hispano motor and engage in commercial flying about the mountains there.[41]

Figure 8. Photograph of a DH-4 aeroplane at Pensacola in 1920. Annotations indicate that it is of MSgt. Ben Belcher, who was the first enlisted Marine Corps pilot. The building visible to the right cannot be positively identified but this may be Old Corry Field. (Courtesy of author)

Figure 9. NAS Pensacola's DH-4B ambulance plane that was stationed at Old Corry Field for emergency transport of injured aviators. (Courtesy National Museum of Naval Aviation)

For emergencies, the U.S. Navy stationed an ambulance plane at Old Corry Field. The ambulance plane, a DH-4B:

...has a stretcher and a doctor's seat in it behind the pilot's cockpit. At our last crash the patient arrived at the dispensary seven miles from the scene of the accident before the news of the crash could be telephoned to the Air Station. The plane made the trip in four minutes.[42]

Unfortunately, in April 1925 the ambulance plane crashed immediately after takeoff, killing its U.S. Marine Corps pilot and seriously injuring a U.S. Navy enlisted man.[43]

NAVAL AIR STATION LANDING FIELD

With a landing field established north of the city, and several miles from the aeronautic station, work had already begun on a landing field a little closer to home.

BOARD OF COUNTY COMMISSIONERS OF ESCAMBIA COUNTY, FLORID

Aviation Field

Figure 10. To this day, the area that used to enclose Old Corry Field is known as "Aviation Field," as this January 2003 rezoning request map indicates. However, very few residents know why it is called Aviation Field. (Courtesy *Pensacola News Journal*)

NOTES

[1] Ellsworth, Lucius & Linda. (1982*). Pensacola The Deep Water City.* Tulsa: Continental Heritage Press, page 66.

[2] Forster, A.E. (N.D.). *World War I, 1917 & 1918.* Pensacola Historical Society.

[3] Flight Testing Newton's Laws. (n.d.). Retrieved from http://www.dfrc.nasa.gov/education/onlineed/newtonslaws/pdf/instructor/.

[4] Hennessy, J.A. (1985). *The United States Army Air Arm.* Washington, DC: Office of Air Force History United States Air Force, page 92.

[5] Sudsbury, E. (1967). *Jackrabbits to Jets.* San Diego, CA: Neyenesch Printers.

[6] Sudsbury, E. (1967). *Jackrabbits to Jets.* San Diego, CA: Neyenesch Printers.

[7] Sudsbury, E. (1967). *Jackrabbits to Jets.* San Diego, CA: Neyenesch Printers, page 59.

[8] Land Planes For Navy. (1922, January 5). *Air Station News, Pensacola, Florida.*

[9] Reports of the Sub Committee on Aviation Fields (1921, December 19). *Blount Folder.* Special Collection, John C. Pace Library.

[10] Land Plane Flying. (1921, December 20). *Air Station News, Pensacola, Florida.*

[11] Blount, F.M.. (Personal Communication, April 22, 1922). *Blount Folder.* Special Collection, John C. Pace Library.

[12] Reports of the Sub Committee on Aviation Fields, (1921, December 19). *Blount Folder.* Special Collection, John C. Pace Library.

[13] Reports of the Sub Committee on Aviation Fields, (1921, December 19). *Blount Folder.* Special Collection, John C. Pace Library.

[14] Admiral Moffett Inspects Sites For Aviation Fields. (1921, October 25). *The Pensacola Journal.*

[15] Admiral Moffett Inspects Sites For Aviation Fields. (1921, October 25). *The Pensacola Journal.*

[16] Admiral Moffett Inspects Sites For Aviation Fields. (1921, October 25). *The Pensacola Journal.*

[17] Admiral Moffett Inspects Sites For Aviation Fields. (1921, October 25). *The Pensacola Journal.*

[18] Reports of Sub Committee on Aviation Fields. (1921, December 19). *Blount Papers.* Special Collection, John C. Pace Library.

[19] Navy To Accept Landing Fields. (1921, December 13). *The Pensacola Journal.*

[20] Trimble, W.F. (1994). *Admiral William A. Moffett.* Washington, DC: Smithsonian Institution Press, page 137.

[21] More Planes Likely For Pensacola Naval Air Station. (1922, June 16). *The Pensacola Journal.*

[22] More Planes Likely For Pensacola Naval Air Station. (1922, June 16). *The Pensacola Journal.*

[23] More Planes Likely For Pensacola Naval Air Station. (1922, June 16). *The Pensacola Journal.*

[24] Kiwanis Field Being Used by Land Planes Recently Added to Naval Aviation Training Course. (1922, July 16). *The Pensacola Journal.*

[25] Christy, H.H. (Personal Communications). (1922, August 3). Condition of Land Field for Land Plane Flying. *Blount Papers.* Special Collection, John C. Pace Library.

[26] Kiwanis Landing Field Ready For Uncle Sam's Use. (1922, October 1). *The Pensacola Journal.*

[27] Old Corry Field Is Dedicated With Impressive Ceremony. (1922, December 8). *The Pensacola Journal.*

[28] Captain Christy's Address At Air Field Dedication. (1922, December 8). *The Pensacola Journal.*

[29] Captain Christy's Address At Air Field Dedication. (1922, December 8). *The Pensacola Journal.*

[30] Old Corry Field Is Accepted – Begin Extensions Soon. (1923, September 9). *The Pensacola Journal.*

[31] Old Corry Field Is Accepted – Begin Extensions Soon. (1923, September 9). *The Pensacola Journal.*

[32] Old Corry Field Completed. (1923, September 20). *Air Station News, Pensacola, Florida.*

[33] Old Corry Field Is Accepted – Begin Extensions Soon. (1923, September 9). *The Pensacola Journal.*

[34] Old Corry Field. (1923, August 5). *Air Station News. Pensacola, Florida.*

[35] Old Corry Field. (1923, August 5). *Air Station News. Pensacola, Florida.*

[36] Pearce, G.F. (1980). *The U.S. Navy in Pensacola.* Pensacola, FL: University of West Florida, page 170.

[37] Barber, B.B. (1924, September 14) Old Corry Field Great Asset To Pensacola; Is Unexcelled In South As Aviation Grounds. *The Pensacola Journal.*

[38] Barber, B.B. (1924, September 14) Old Corry Field Great Asset To Pensacola; Is Unexcelled In South As Aviation Grounds. *The Pensacola Journal.*

[39] Class 21 Nears Completion of Seaplane Tests. (1925, March 4). *The Pensacola Journal.*

[40] Old Corry Field Is Accepted – Begin Extensions Soon. (1923, September 9). *The Pensacola Journal.*

[41] Gypsy Ride By Station Pilots. (1923, September 5). *Air Station News, Pensacola, Florida.*

[42] Raby, J.J. (1924, December 28). Naval Aviation Training. *The Pensacola Journal.*

[43] Marine Flier Killed. (1925, April 3). *The Pensacola Journal.*

CHAPTER 16
STATION FIELD

Figure 1. DH-4B over Station Field at Naval Air Station Pensacola in August 1922. Notice the two dirigible hangars in the lower right. As noted by Reginald Wright, author of *Contact!*, the former floating dirigible hangar is now on "shore duty." (Courtesy U.S. Navy)

BLIMP FIELD TO LANDING FIELD

In December 1921, shortly after the Secretary of the Navy accepted Pensacola's offer of Old Corry Field, an announcement was made at the Naval Air Station that $18,000 had been made available for "...improving and enlarging the 'blimp' field;[a] and from there land flying operations will function. The field will be made longer by extension along the line of the prevailing winds, thus allowing ample space for landing and take offs."[1] Naval authorities also announced that eight landplanes would be transferred to Pensacola to begin landplane training. Delivery might be as early as January 1922 and the station's initial complement was expected to be sizeable, totaling nearly 36 JN-6s, Voughts, and DH-4s.[2]

Therefore, in preparation for flight operations aboard the USS LANGLEY, Naval Air Station Pensacola was about to begin its first formal course of instruction in landplanes. Not surprisingly, given the competitive nature of U.S. Army and U.S. Navy relationships, the immediate concern expressed by

---

[a] NAS Pensacola had stopped all lighter-than-air training in August 1921.

Pensacola Naval aviators was that the U.S. Army would conduct this training. However, these rumors were quickly dispelled when Washington announced that the U.S. Army would not be associated with this new training in any way.[3]

NAS Pensacola's *Air Station News* expressed the reasoning behind the U.S. Navy conducting its own landplane training instead of the U.S. Army:

In the past, students who were selected for training in land planes were sent direct to the army fields without any instruction or tests whatsoever, to determine their qualifications for advanced instruction in land machine flying. As a consequence considerable time and money was spent on a student before it was determined whether or not he was fitted for that branch of flying. With means for giving preliminary instruction here on this station, however, it can be determined at once whether or not a student is adapted for scout combat work and land spotting – the two important courses in the land flying course.[4]

STATION FIELD

Figure 2. The first landplanes at NAS Pensacola were flown from the old balloon field north of the station. (Courtesy of National Museum of Naval Aviation)

The site selected by the U.S. Navy for its first landing field in Pensacola was the blimp field located immediately north and outside of the station's brick walls.   Eventually, $27,000 was allocated to prepare the blimp field for landplane operations and a landing strip 300 feet wide by 900 feet long was leveled and cleared of all timber, stumps and undergrowth.[5]

On February 21, 1922, construction began, although progress apparently proceeded slowly as evidenced by this March report:

Only two teams are at work on the excavation and leveling, due to the fact that the machinery for the work has not been received on this Station. Ten scrapers, ordered and due on the 23rd, have not yet arrived, altho they are expected daily. When they arrive, ten teams will be put on this job and work will be expedited, with the intention of completing operations at the earliest possible date. It is hoped to have the dirigible field in condition for land plane operations by the time the land planes which are on the way to Pensacola arrive at this Station.[6]

Station Field, as it soon became known, covered 25 acres when first completed. In June 1922, a contract was let to "...construct 44,000 square yards of hard-surfacing similar to that part of the landing field already built. Clay is to be used and the job is to be completed in 150 days."[7] In July 1925 further construction expanded Station Field to 62 acres when "trees and stumps to the north and northwest have been removed and the cleared space has been leveled and filled in."[8]

EARLY LANDPLANES IN PENSACOLA

Landplanes were not entirely foreign to Pensacola in the early 1900s. A hydroaeroplane capable of conversion to a landplane was part of the complement of planes that arrived with the first Naval and Marine aviators in early 1914.

There is also evidence that as early as September 1914 landplanes were arriving in Pensacola for testing. *The Pensacola Journal* described one such plane.

A new airship entirely different it is said from any which have been in use at the Pensacola aeronautical station arrived here this week and was delivered at the local station on Tuesday last. The new aircraft was fitted with wheels so that it is possible of navigation on level ground with apparently as much ease as the airboats.

Following the arrival of the new craft, the announcement is semi-officially made that several additional aviators are to be on duty soon at the Pensacola station. It was said that four more aviators will shortly arrive for this point.

Further, in December 1916, Lieutenant P. Rader, a demonstrator for the Curtiss Company, told *The Pensacola Journal* that a land machine might be sent to Pensacola instead of their factory for testing as "...every facility is to be had for the maneuvers" the aeroplane needs.

A worker described the field in operation:

No concrete runways, it was all dirt... they started using a lot of red clay that they brought in. First they brought it in on flat cars and it had to be picked up and hauled to where they wanted it and then at a later date they probably made a contract with somebody to have it brought in big trucks. And truck load after truck load of it came in for a long, long time. They would get in and spread it around on the field. The wind would blow and blow. Every way the wind blew that was the way this clay would blow. If it blew from the south, we got it at the big Bayou. If it blew from the east then the officers and their homes got it. And it was a terrible thing for a long time.

Figure 3. DH-4 flying over NAS Pensacola's Station Field. May 1927 was the last month of use for DH-4s in the U.S. Navy. (Courtesy National Museum of Naval Aviation)

Now they started using these Curtiss Jennies on the field and they had a steel end on it something that would slide along the bottom and dig into the ground. And it was used to steer the planes on the ground. And every time they started to steer a plane on the ground there would be a cloud of clay go into the air and which ever way the wind blew that clay went. Well that went on for some time, for a long time. I think the clay eventually, they were able to keep it dampened, kept it so it wasn't too bad or at least it wasn't as bad as it had been.[9]

STATION FIELD RENAMED CHEVALIER FIELD

The first landing field at Naval Air Station Pensacola was known simply as "Station Field." Located north of the old Pensacola Navy Yard's brick walls, it was constructed in 1922 upon a balloon field built on the site of the former village of Woolsey.

In 1936, apparently without fanfare, Station Field was renamed "Chevalier Field" in honor of Lieutenant Commander Godfrey de Courcelles Chevalier, Naval Aviation Number 7, who died November 14, 1922 from injuries sustained after an aeroplane crash at Norfolk, Virginia.

Chevalier Field served as NAS Pensacola's primary landing field until the arrival of the first jets. With a maximum runway length of less than 4,000 feet, Chevalier Field could not accommodate the longer runway requirements of these modern aircraft.

To support jet operations, Sherman Field, current home to the Blue Angels and Training Air Wing SIX was built and officially dedicated on November 2, 1951.

Chevalier Field was then used primarily for test flights from the Naval Aviation Depot, tasked with the overhaul and repair of fleet aeroplanes. On November 1, 1965 Chevalier Field was officially closed. It is now home to the schoolhouses of the Naval Air Technical Training Center.

LANDPLANE TRAINING COMMENCES

The *Air Station News* reported in February 1922 that "...fifty JNs are to be shipped to this Station from the Army depot at Americus, Georgia. The bill of lading for this shipment has not been received, but it is expected in the near future."[10] Additionally, it reported that "A shipment of Voughts, DHs and JNs for the purpose of training pilots and making tests is under way and should reach the Station here soon."[11] No record could be found of when the landplanes finally arrived in Pensacola, but it is unlikely they were flown down. They were probably crated, shipped by train, and unboxed and erected at the station.

The first class of flight students to be trained in landplanes began the five-week course in July 1922. It was noted that the "large clay-surfaced landing field is not yet completed but this is in no way hindering the officers in completing their course."[12]

On August 9, 1922 the first seven Naval aviators successfully completed the landplane training course.[13] Flying JN-4-H type landplanes they were noted to be "Fully qualified in aerobatics, precision landings and cross country work."[14] The second class in landplane instruction started August 29, 1922.[15]

Pensacola.—Station Field, Navy. Immediately N. of U.S. Naval Air Station; 4½ miles SW. of Pensacola. Altitude, 30 feet. Rectangular, 3,600 by 2,200 feet with sand asphalt taxiway 100 feet wide along N. and E. sides of field; sand and clay, level, natural drainage; "USN PENSACOLA" on hangar. Landing area flood lights. Facilities for servicing Government aircraft only. Seaplane anchorage adjacent to S. side of field, landing and take-off in Pensacola Bay. Mooring buoys and ramps available.

Figure 4. 1934 U.S. Bureau of Commerce description of Station Field. (Courtesy of Chris Kennedy)

It appears that Station Field's landing field environs were not entirely safe. One Naval aviator remembered:

In January 1923, the Langley went to Pensacola. The air station there had just inaugurated the first land plane classes and had a small field alongside the hangar for lighter-than-air craft. The trolley line to town passed alongside one side of the field and the approach over the trolley wire was a mental hazard.[16]

To lessen this danger, the Pensacola Electric Company, which provided transportation from the City of Pensacola to the aeronautic station, relocated their system away from the landing field.[17] An article noted:

With the removal of the carline and the high tension line from its present location to the proposed route the hazards to planes taking off in a northwesterly direction will be removed to a great extent. Especially is this so in the case of student aviators who are not yet fully qualified in landplanes. Under present conditions, flying from the field when the wind is from the north, northwest, or west is considered extremely dangerous.[18]

CROSS COUNTRY FLIGHTS

Naval aviators would not take long before flying their landplanes out into the Pensacola countryside. The *Air Station News* reported that on August 8, 1922 "Five of the JN-4-H land planes attached to Squadron Six made a cross country flight to Montgomery, Alabama...the distance of approximately one hundred and fifty miles was covered in two laps, a stop being made at Andalusia, Alabama for gasoline and oil."[19]

In one unprecedented journey in August 1922, Lieutenant T.B. Lee and Lieutenant Elliott flew two DH-4Bs from Pensacola to Washington. The trip took "...nine hours and fifty-five minutes, including the time lost looking for landing fields at Ellaville and Columbia."[20] The only problem was that this flight was spread over several days due to engine problems and landing accidents.

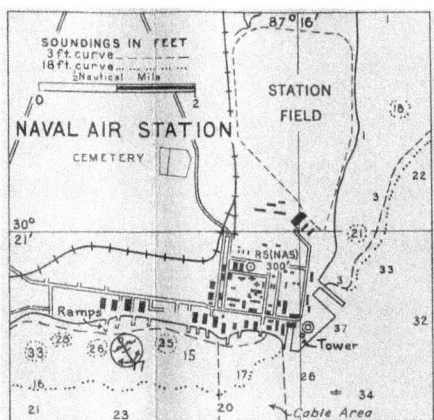

Figure 5. Aviation chart showing Station Field's position relative to the air station and seaplane handling facilities. (Courtesy of Chris Kennedy)

Figure 6. Dwarfed by the two dirigible hangars, precisely aligned landplanes sit on the flight line of Station Field. (Courtesy National Museum of Naval Aviation)

This time would be shortened on April 17, 1923. Lieutenant Commander C.P. Mason, USN and Lieutenant H. J. Norton, USMC "...made the first successful one-day flight from Pensacola to Washington."[21] Piloting a DH-4B, they departed Old Corry Field at 4:55 a.m. and landed in Washington at 6:00 p.m. with only a single stop at Columbia, South Carolina. Captain H.H. Christy, commandant of the Naval Air Station, stated that "This is the first time that such a flight has been made in a single day. The flight was highly satisfactory as an aviation accomplishment."[22]

Figure 7. The introduction of jets into Naval aviation, shown here flying south of NAS Pensacola, foretold the end of Chevalier Field. (Courtesy National Museum of Naval Aviation)

This achievement also presented an opportunity for the U.S. Navy to brag about Pensacola. The *Air Station News* noted:

Tuesday's accomplishment demonstrates the value of Pensacola's geographical location as an aviation center. Any officer on the station could be hastily summoned to Washington for important consultations or other work and make the trip in a single day. If the navy's aviation activities were concentrated on the Pacific coast or some other far distant point this would be utterly impossible.[23]

Apparently, the first time Pensacolians observed the new landplanes up close was a Navy Day celebration on October 27, 1922. The *Air Station News* reported:

Three JN land planes will stunt over the football field either before the football game or between halves. This is the first time JNs have been seen in stunts by the public since the opening of the land plane school at this Station. These planes flown by expert pilots, will engage in all types of stunts, and this feature of the day's exercises promises to be of great interest.[24]

FIRST LANDPLANE CRASH

The first landplane mishap at Station Field occurred on December 14, 1922 when a "...crash resulted from a side slip upon the landing which brought the plane down in the shallow water near the landing field."[25]

SATURATION AND DANGER

However, before the first spade of sand had been shoveled on Station Field, the U.S. Navy had recognized that the old blimp field was too small to handle the expected large training demands for landplane pilots. Further, with so many landplanes and seaplanes flying in close proximity over Pensacola Bay, flight operations were "not only unsatisfactory but actually dangerous."[26]

Therefore, Old Corry Field assumed the greater burden of landplane training. However, Pensacola had only leased this landing field, not purchased it. And the leases were about to expire.

NOTES

[1] Land Plane Flying. (1921, December 20). *Air Station News, Pensacola, Florida.*

[2] Land Plane Flying. (1921, December 20). *Air Station News, Pensacola, Florida.*

[3] Allotment for Landing Field Has Been Received. (1922, February 5). *Air Station News, Pensacola, Florida.*

[4] Land Plane Flying. (1921, December 20). *Air Station News, Pensacola, Florida.*

[5] Allotment For Landing Field Has Been Received. (1922, February 5. *Air Station News, Pensacola, Florida.*

[6] Progress On Flying Fields For The Land Planes. (1922, March 5). *Air Station News, Pensacola, Florida.*

[7] Air Station Will Get Improvement. (1922, June 20). *The Pensacola Journal.*

[8] Landing Field At Air Station Being Enlarged. (1925, July 21). *The Pensacola Journal.*

[9] Forster, A.E. (N.D.). *Story of Lindbergh.* Pensacola Historical Society.

[10] New Planes For Pensacola. (1922, February 20). *Air Station News, Pensacola, Florida.*

[11] New Planes For Pensacola. (1922, February 20). *Air Station News, Pensacola, Florida.*

[12] Twenty Officers Fly Land Planes. (1922, July 11). *The Pensacola Journal.*

[13] Qualify In Land Planes. (1922, August 20). *Air Station News, Pensacola, Florida.*

[14] Qualify In Land Planes. (1922, August 20). *Air Station News, Pensacola, Florida.*

[15] Second Land Plane Class. (1922, September 5). *Air Station News, Pensacola, Florida.*

[16] Tate, J.R. (1998). We Rode the Covered Wagon. In Wooldridge, E.T. (Ed.) *The Golden Age Remembered.* Annapolis, MD: Naval Institute Press, page 71.

[17] Made Nearly 400 Flights At Navy Yard During Week. (1924, May 15). *The Pensacola Journal.*

[18] Flying Hazards at Navy Landing Field Lessoned. (1924, March 29). *The Pensacola Journal.*

[19] Cross Country Flying. (1922, August 20). *Air Station News, Pensacola, Florida.*

[20] Flight of DH-4-B Planes To Washington, D.C. (20 August 1922). *Air Station News, Pensacola, Florida.*

[21] Lieutenant Commander Mason On One Day Hop To Washington. (1923, April 20). *Air Station News, Pensacola, Florida.*

[22] Lieutenant Commander Mason On One Day Hop To Washington. (1923, April 20). *Air Station News, Pensacola, Florida.*

[23] Lieutenant Commander Mason On One Day Hop To Washington. (1923, April 20). *Air Station News, Pensacola, Florida.*

[24] 1922=Navy Day=1922. (1922, October 25). *Air Station News, Pensacola, Florida.*

[25] Land Plane Crash. (1922, December 20). *Air Station News, Pensacola, Florida.*

[26] Evolution of Corry Field. (1935, October 28). *Air Station News, Pensacola, Florida.*

CHAPTER 17
NEW CORRY FIELD

Figure 1. View of New Corry Field in July 1956. (Courtesy National Museum of Naval Aviation)

LEASES EXPIRE

The leases acquired by the Chamber of Commerce for Old Corry Field and sub-leased to the government were for only five years, expiring on July 1, 1927. Civic leaders feared that once these leases expired, the U.S. Navy might be enticed to move their landplane training elsewhere. This sentiment was captured in a front-page newspaper article:

...another thought that has gained public credence is the likelihood of some other section becoming a stronger bidder for the location of a naval flying field, and that there being no investment, holdings or hindrance regarding Old Corry Field, Pensacola is not as secure as she might be if the field actually belonged to the government. A number of citizens ever interested in Pensacola's future and its continued forward progress have interested themselves within the past few weeks in evolving a plan by which this one of the city's advantages might become a permanent fixture.[1]

This article expressed most people's opinion that the best course of action would be to purchase Old Corry Field outright and turn it over to the U.S. Navy, avoiding any lease renewal issues. The U.S. Navy would need authorization to accept such land. Therefore, in early 1925 Senator Duncan U. Fletcher introduced a resolution passed by the Senate that authorized the Secretary of the Navy to acquire title to land suitable for landplane training.[2]

The actual purchase of the land was simplified by the fact that "The leases obtained from the owners of the property contained an option of purchase at a cost of approximately $56,000 at any time up to July 1, 1927."[3] In August 1926, county commissioners "...adopted a resolution pledging themselves to take such steps as they legally could whereby the county will purchase the field and extend its use to the government for an indefinite period."[4] A short time later *The Pensacola Journal* reported "Escambia county will take immediate steps to purchase Old Corry Field and turn it over to the navy department to insure the maintenance here of the naval air station."[5]

Admiral Moffett was delighted with this plan and sent a note in August 1926 that stated in part:

...the bureau favors the continuance of primary land plane training at Pensacola owing to the many advantages that the Naval Air Station and Old Corry Field possess and is recommending to the Secretary of the Navy that primary land plan training be continued after July 1, 1927 at Pensacola.[6]

Meanwhile landplane training in the U.S. Navy was expanding at an exponential rate, including an announcement by the U.S. Navy that it expected a "...fifty percent increase in planes and equipment at the Pensacola Naval Air Station."[7] Consequently, the U.S. Navy began to hint that they desired a larger landing site than the present Old Corry Field, preferably one that was owned and not leased. Again, civic leaders needed to act quickly if Pensacola was to keep Naval aviation training.

PROBLEMS EXPANDING OLD CORRY FIELD

The simple solution was to expand Old Corry Field to meet the U.S. Navy's needs. While options to buy on the current leases of Old Corry Field were available, the Bureau of Aeronautics had requested that the field be increased in size to at least 500 acres. Purchase prices of the adjacent 250 acres were hampered by the fact that "...the owners of this property were demanding a price more than the land was worth"[8] and "...at a figure which local citizens deemed entirely too high making it prohibitive."[9]

Further, Old Corry Field was distant from the Naval Air Station and the City of Pensacola had begun to encroach around the field "Endangering both the aviators who flew from the field and the civilians living around it."[10]

The issue of expanding Old Corry Field soon became moot. The Bureau of Aeronautics let it be known that "...a site nearer than Corry Field to the air station would be more acceptable than the site at Corry Field, provided that an area of 500 acres could be obtained."[11] This new field should also be "So located that extension of the city would not encroach upon it."[12]

In less than two and a half years, the U.S. Navy had outgrown Old Corry Field. Once again, Naval authorities began to look at other Naval air stations to absorb landplane training. However, the civic leaders of Pensacola were not about to risk losing Naval aviation. They had a plan.

For the second time, Pensacola civic leaders formed a joint committee to search for a suitable landing field site. The U.S. Navy had requested a permanent site of "...at least 500 acres lying in a rectangle and of suitable soil and practically level, conveniently located with respect to the Air Station, and accessible thereto by highway and railroad, where water could be secured..."[13]

In February 1927, eight possible landing field sites were identified and submitted to the commandant of the Naval Air Station for consideration.[14] One site located "...west of Pensacola near the Jackson road, and about a mile north of the Gulf Beach highway"[15] offered several advantages over the present Old Corry Field:

...it has a much larger area; it is about 5 miles nearer the naval air station and thus will greatly lessen transportation and communication difficulties; its location will eliminate necessity for personnel to fly over the city of Pensacola to reach it from the naval air station; and it permits relatively close concentration of station activities.[16]

A Congressional report advanced strong arguments for retaining landplane training at Pensacola and selection of the proposed site by the U.S. Navy:

The men are sent to Pensacola for seaplane training. Landplane training should follow immediately thereafter, and should the landplanes be at some other location, there would be considerable expense in transporting these men; shops, storehouse facilities, quarters and barracks are sufficient at Pensacola to provide for these men undergoing landplane training. Owing to the fact that the Prieto tract is so close to the naval air station, the bureau does not contemplate any extensive development to this field, as

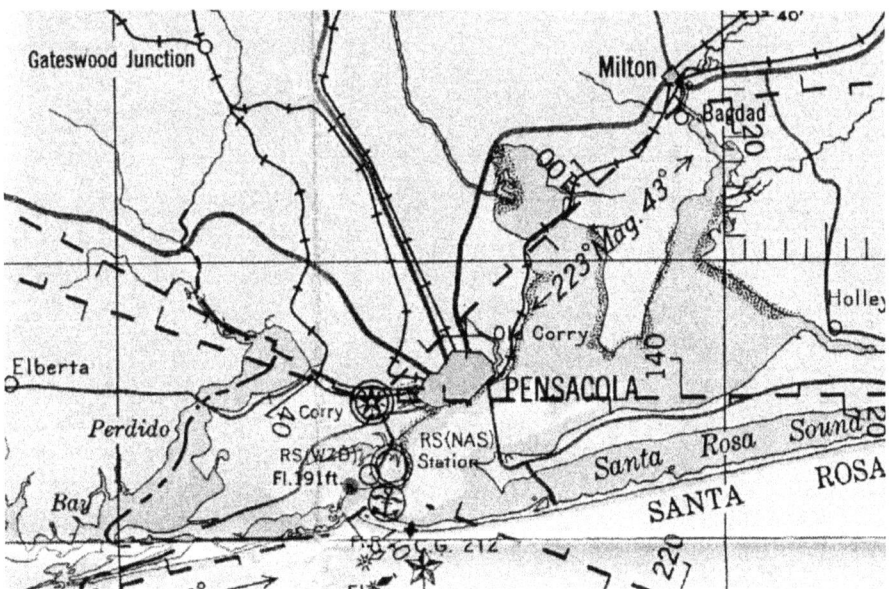

Figure 2. This insert from a 1934 aviation map illustrates the various military landing sites available for transient aviators: the seaplane base at the Naval Air Station, Station Field, New Corry Field and Old Corry Field. Auxiliary landing fields or civilian fields such as Pensacola Municipal Airport are not shown. (Courtesy of Chris Kennedy)

PENSACOLA (NAVAL AIR STATION), FLA.

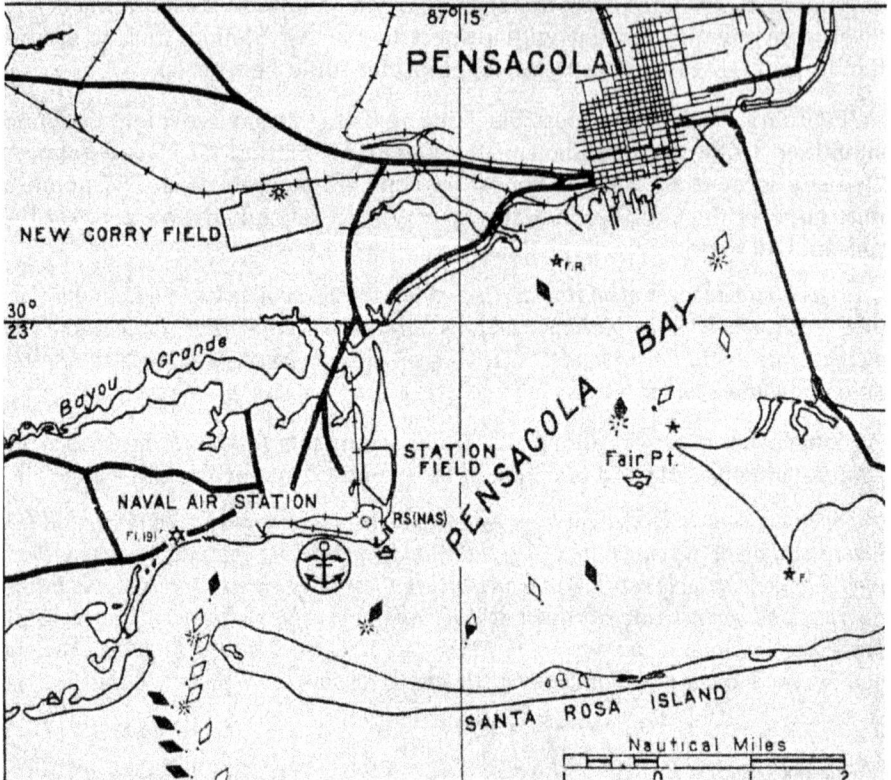

Figure 3. A more detailed insert from 1935 showing New Corry Field in relation to the Naval Air Station and the City of Pensacola. This field was closed in 1958 and is currently home to the Naval Technical Training Center. (Courtesy of Chris Kennedy)

the station will be used for all major repair work on planes and as the main center of activity for the training. Some hangars may be put on the field to house the planes during the training period to avoid flying back to the station each night. Climatic and general flying conditions at Pensacola are excellent, and the bureau desires to retain the landplane training at this location.[17]

The U.S. Navy agreed. Escambia County then purchased the new site, issuing $100,000 in bonds.[18a] The Chamber of Commerce also agreed to "...place 250 acres of the tract in condition for flying, construct a railroad spur to the site, and construct and keep in repair a hard-surfaced road from the west Pensacola road to the site."[19]

A bill was introduced by Senator Duncan U. Fletcher and Representative J.H. Smithwick that "...the secretary of the navy be authorized to accept 500 acres of land west of Pensacola for use as a new aviation field, in connection with

---

[a] Actually, only $69,300 in bonds were sold and they were paid off in 1935.

the United States Naval Air Station here."[20] The government soon approved the offer and bids were opened in July 1927 by the Chamber of Commerce to clear and grade 250 acres of the field.[21] Eventually, deeds for the 500 acres that comprised the new landing field were turned over to the Navy Department.[22]

Was the price of obtaining and improving Corry Field worth it? Within two years the U.S. Navy announced that more than $5,000,000 would be invested in Pensacola Naval Air Station. An editorial in *The Pensacola Journal* stated "The lesson that Pensacola may learn from the investment of $100,000 in the Naval Air Station, is one that may be carried into all development here: You can't get something for nothing. Pensacola must give as well as take."[23]

NAVAL AVIATION SAVED AGAIN

When Rear Admiral Moffett next visited Pensacola, he made a speech reminiscent of his visit five years previous after Old Corry Field had been presented to the government. He told an assembled luncheon crowd at the San Carlos Hotel that Pensacolians had "...'builded [sic] better than they knew' when they purchased the 509 acre Praeto [sic] grant and presented it to the Navy Department as a permanent land plane training field..."[24]

The Admiral went on to say that "...it had been decided...to transfer a major portion of the activities from this station to San Diego, but the transfer was delayed for several reasons. Had the move been made...the training activities here would have been virtually abandoned."[25]

CONSTRUCTION AND IMPROVEMENTS

Construction at the new field proceeded rapidly. In March 1927 it was announced that:

Bids on the clearing, grading and conditioning of the new landing field will be opened on March 18th, and work undertaken shortly thereafter, to insure the 250 acres necessary being in shape by July first, at which time Corry Field will be abandoned. Additional work involved is the construction of a spur track to the field, the building of road, and the transfer of buildings, gasoline stowage and the like, from Corry Field.

The approaches to the new field are on the whole all fairly good. From the South planes come in over "Blackjacks," and the approach will be otherwise clear when permission

---

Pensacola.—New Corry Field, Navy. Four miles W. of Pensacola, 4 miles N. of U.S. Naval Air Station. Altitude, 35 feet. Rectangular, 5,850 feet E./W., 3,000 feet N./S., sandy loam, level, natural and artificial drainage. Taxiway around field and through center. U.S.N. on hangar, illuminated. Surrounded by trees; hangars and administration building on N., high water tower and stack to N., buildings and gasoline storage tanks in NE. corner. Beacon, boundary, obstruction, and landing area flood lights. U.S.N., boundary, obstruction, and landing area flood lights not turned on except upon notice of arrival of aircraft. Beacon, 24-inch clear, with auxiliary code beacon flashing characteristic "N" (— .), operates from sunset to sunrise. Facilities for servicing Government aircraft, day and night; other planes in emergency.

Figure 4. 1934 U.S. Bureau of Commerce description of New Corry Field. (Courtesy of Chris Kennedy)

has been granted to cut down some fifty pine trees scattered through this area. The East and West approaches are unobstructed, as is also the North, with the exception of the Northeast corner, which is fairly heavily wooded. [26]

A local humorist noted that the feat of clearing of the landing area was "… accomplished by removing four billion and nine hundred million of the small, delicate shrubs, classically called 'The Sailor's Friend,' but known locally to the natives as 'Blackjacks.'" [27]

Another gentleman noted that:

The field was only half cleared and was at best merely a level area clear of blackjacks with a few sparsely scattered frame buildings in one corner. The duty officer could frequently be found after working hours strolling about the field with a shot gun on his shoulder hunting quail and rabbits. And in the center of the western end was a marshy pond which abounded in snipe, and an occasional duck would also drop in for serving. [28]

Nonetheless, in June, 1927 a Naval officer speaking at some railway hearings testified "That the United States government intends to spend a half million dollars on buildings and improvements at the new naval landing field west of Pensacola within the next five years" and that "…the new field will be the third largest in the entire United States." [29]

The U.S. Navy also began dismantling all of the structures at Old Corry Field for rebuilding at New Corry Field. [30] Construction of new barracks capable of housing more 100 men, mess halls, galley, office buildings, gasoline tanks, and hangars began and were designed to meet a July 1, 1927 opening coincident with the expiration of the leases at Old Corry Field. [31]

Figure 5. Overgrown after Corry Field was closed, these airfield lights stood silently in the woods for decades until revealed by road expansion in 2000. (Courtesy of author)

To avoid trespassers of both the two and four-legged kind, "The cleared area of 250 acres is being fenced in as a protection against cattle and the use of the old trails running through the field." [32] Also "An ambitious bootlegger had set a still nearby to furnish the personnel with some anti-flying juice, but through the efforts of the prohibition squad from Pensacola and Captain Slatton of the Yard Police, his services were dispensed with – at least, his still and equipment are now the property of the County." [33]

Actual flight operations at the new site began on July 5, 1927 a few days in advance of the formal opening of the landing field. [34]

NEW CORRY FIELD DEDICATED

The Navy Department announced that the name of Corry Field would be used by the new landing field.[35] The official dedication ceremony was simple and occurred on July 9, 1927 with the raising of the United States flag promptly at 10 o'clock.[36] Rear Admiral F.B. Upham, Commandant of the Naval Air Station, escorted a host of local dignitaries on a tour of the new buildings and crossed the field to stop in front of fifty planes.

As described by The *Pensacola News Journal*:

Lined up abreast, the planes, representing an expenditure of $85,000, presented a magnificent picture, with their pilots and mechanicians standing by 'ready to ride.' And ride they did just as soon as the commandant's inspection had been completed. Twenty of the planes took off at intervals of 15 seconds, circled the field and later 'did their stuff' for the edification of the onlookers.[37]

RETURN TO OLD CORRY FIELD

Not everything went smoothly at the new landing field. In February 1928 it was decided that "...rains have made it difficult to operate planes on some sections of new Corry Field..." and permission was granted from owners of Old Corry Field to operate from there.[38]

In July 1929, the U.S. Navy returned to Old Corry Field on a permanent basis.[39] A 100-acre tract was leased from Nettie F. Ripple to be used for student flight training.[40]

The reasons for the return of the Naval aviators to Old Corry Field were the same reason the U.S. Navy had left two years before:

The landing fields at New Corry Field and the Station are at times crowded since there has been an increase in the number of students, so that additional landing space is needed. Besides, both in the routine of training and in case the engine or plane requires minor attention, there are times when it is more satisfactory to land at 'Old Corry' than to return to one of the other fields.[41]

The U.S. Navy asked the city for minimal facilities and the Chamber of Commerce agreed "...to erect an observation tower, where a Hospital Corpsman and an Aviation Machinist's Mate will be on duty. The tower will be equipped with two telephones, one on a wire direct to the Air Station and the other connected with the Pensacola exchange."[42]

ZEE FIELD

The first hint of a planned auxiliary landing field for Pensacola's Naval aviators was a brief article in the station newspaper in September 1929:

The Bureau of Aeronautics has authorized the Commandant to investigate the availability of a suitable tract of land to be leased for use as an auxiliary landing field. An additional field has become necessary because of the congested condition of land plane flying at Corry Field. The proposed field will be in addition to old Corry Field. Negotiations for a tract of land are well under way.[43]

Lease negotiations must have proceeded smoothly and rather swiftly for NAS Pensacola's new official auxiliary landing field. Known as Zee Field, less than three months after negotiations began it was called upon to serve as a divert field when heavy fog rapidly rolled into Pensacola, forcing over 37 aeroplanes to seek alternative landing sites.[44]

For Pensacolians, with three military landing fields and one military auxiliary landing field in operation, momentum for a civilian landing field to call their own was slowly building.

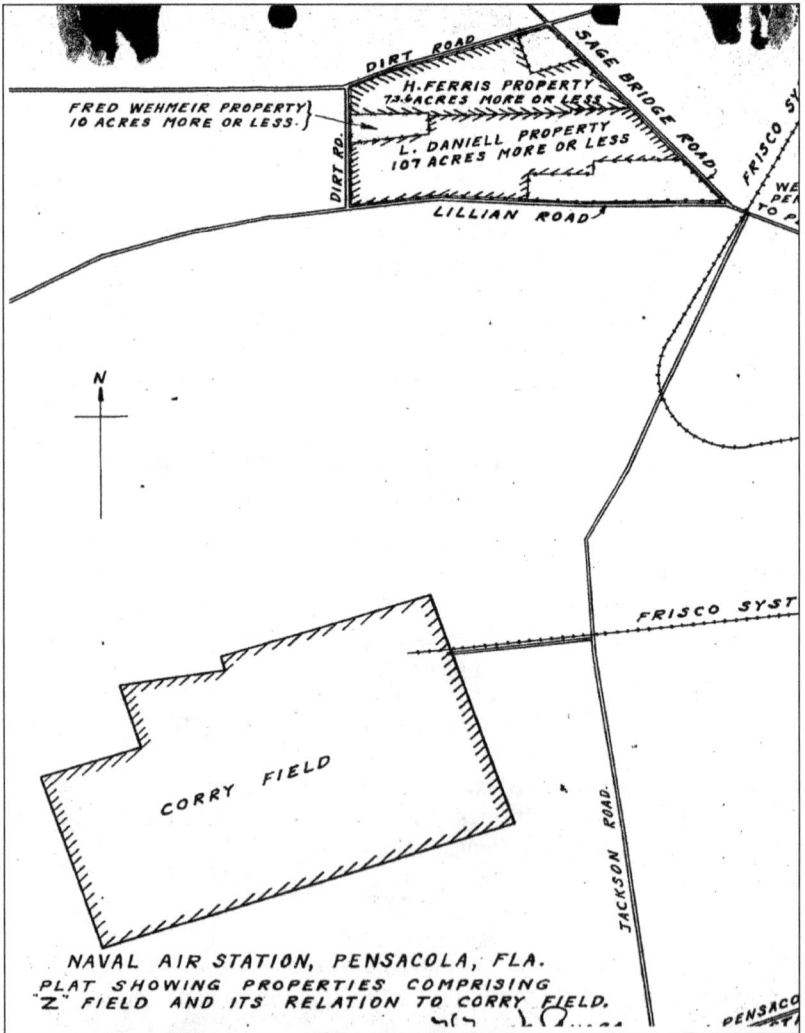

Figure 6. As shown on this plat, Zee Field was located on leased land north of New Corry Field "near the Lillian Bridge road a mile or so beyond the Frisco railroad." (Courtesy of CDR Doug Siegfried, USN (Ret))

NEW CORRY FIELD CLOSED

Overrun by residential and commercial growth, and situated too close to Sherman Field and Saufley Field for safe flight operations, Naval Auxiliary Air Station Corry Field was closed in 1958. The introduction of jets into Naval aviation and subsequent reduction in requirements for propeller-trained pilots also had doomed the U.S. Navy's first auxiliary field.

Initially, the City of Pensacola thought to acquire the field to establish a college, reportedly at an asking price of $13.5 million, but the U.S. Navy retained the field. In 1960, communications technicians, later designated cryptologic technicians, were assigned to Corry Field for instruction. Hangars and other buildings were converted into classrooms as Corry Field transitioned from flight training to academic training. In 1973, the field was redesignated as Naval Technical Training Center Corry Field, as it is known today.

NOTES

[1] Chamber Takes Initial Steps for Purchase Of Old Corry Field For Navy. (1926, August 18). *The Pensacola Journal*.

[2] Navy May Purchase Land Training Field Here. (1925, January 25). *The Pensacola Journal*.

[3] Authorizing Acceptance By Navy Department Of A Site For An Aviation Training Field In The Vicinity Of Pensacola, Fla., Report No. 2038, 69th Cong. (1927).

[4] Plan To Save Old Corry Field For Naval Flying Trainees. (1926, August 12). *The Pensacola Journal*.

[5] County To Buy Old Corry Field For U.S. Navy. (1926, August 25). *The Pensacola Journal*.

[6] Navy Accepts Old Corry Field For Trainees. (1926, September 2). *The Pensacola Journal*.

[7] Pensacola's Pride. (1925, July 26). *The Pensacola Journal*.

[8] Authorizing Acceptance By Navy Department Of A Site For An Aviation Training Field In The Vicinity Of Pensacola, Fla., Report No. 2038, 69th Cong. (1927).

[9] New Flying Field Offered Government. (1927, February 8). *The Pensacola Journal*.

[10] Evolution of Corry Field. (1935, October 28). *Air Station News, Pensacola, Florida*.

[11] Authorizing Acceptance By Navy Department Of A Site For An Aviation Training Field In The Vicinity Of Pensacola, Fla., Report No. 2038, 69th Cong. (1927).

[12] Evolution of Corry Field. (1935, October 28). *Air Station News, Pensacola, Florida*.

[13] New Flying Field Offered Government. (1927, February 8). *The Pensacola Journal*.

[14] New Flying Field Offered Government. (1927, February 8). *The Pensacola Journal*.

[15] Upham Thanked for His Work On New Field. (1927, July 6). *The Pensacola Journal*.

[16] Authorizing Acceptance By Navy Department Of A Site For An Aviation Training Field In The Vicinity Of Pensacola, Fla., Report No. 2038, 69th Cong. (1927).

[17] Authorizing Acceptance By Navy Department Of A Site For An Aviation Training Field In The Vicinity Of Pensacola, Fla., Report No. 2038, 69th Cong. (1927).

[18] Corry Field Bond Issue Is Provided For. (1927, July 27). *The Pensacola Journal*.

[19] Authorizing Acceptance By Navy Department Of A Site For An Aviation Training Field In The Vicinity Of Pensacola, Fla., Report No. 2038, 69th Cong. (1927).

[20] New Flying Field Offered Government. (1927, February 8). *The Pensacola Journal.*

[21] To Open Bids For Air Field. (1927, March 11). *The Pensacola Journal.*

[22] Corry Field Be Used Soon. (1927, July 2). *The Pensacola Journal.*

[23] Giving And Taking. (1929, July 23). *The Pensacola Journal.*

[24] Naval Air Station Activities Here To Be Expanded. (1927, April 27). *The Pensacola Journal.*

[25] Naval Air Station Activities Here To Be Expanded. (1927, April 27). *The Pensacola Journal.*

.[26] New Landing Field. (1927, March 20). *Air Station News, Pensacola, Florida.*

[27] Corry Field, 'The Voice Of The Wilderness'. (1932, October 5). *Air Station News, Pensacola, Florida.*

[28] Dedication of Corry Field. (1935, January 1). *Air Station News, Pensacola, Florida.*

[29] Government To Spend Half Million On Landing Field. (1927, June 17). *The Pensacola Journal.*

[30] New Barracks Being Erected. (1927, May 6). *The Pensacola Journal.*

[31] Work On Landing Field Is Progressing Rapidly. (1927, June 19). *The Pensacola Journal.*

[32] Work On Landing Field Is Progressing Rapidly. (1927, June 19). *The Pensacola Journal.*

[33] Corry Field In Commission. (1927, July 5). *Air Station News, Pensacola, Florida.*

[34] Corry Field Be Used Soon. (1927, July 2). *The Pensacola Journal.*

[35] Name Of Corry Will Be Kept By New Field. (1927, May 1). *The Pensacola Journal.*

[36] New Corry Field Officially Given To Government. (1927, July 10). *The Pensacola Journal.*

[37] New Corry Field Officially Given To Government. (1927, July 10). *The Pensacola Journal.*

[38] Navy Aviators Use Old Field. (1928, February 14). *The Pensacola Journal.*

[39] Navy Secures Landing Field. (1929, July 4). *The Pensacola Journal.*

[40] Old Corry To Be Re-Commissioned. (1929, July 20). *Air Station News, Pensacola, Florida.*

[41] Old Corry To Be Re-Commissioned. (1929, July 20). *Air Station News, Pensacola, Florida.*

[42] Old Corry To Be Re-Commissioned. (1929, July 20). *Air Station News, Pensacola, Florida.*

[43] New Landing Field. (1929, September 5). *Air Station News, Pensacola, Florida.*

[44] Fog Bound. (1929, December 20). *Air Station News, Pensacola, Florida.*

CHAPTER 18
PENSACOLA MUNICIPAL AIRPORT

Figure 1. 1935 photograph of the first hangar at Pensacola Municipal Airport. (Courtesy of C. H. Blanchard, Jr.)

CALLS FOR A COMMERCIAL LANDING FIELD IN PENSACOLA

During his visit in May 1927 to celebrate Pensacola's acquisition of New Corry Field for the U.S. Navy, Rear Admiral Moffett made a suggestion. In view of the rapid growth of commercial aviation across the nation, he recommended that Pensacola establish a commercial landing field and cited "…(Old) Corry Field as an ideal site for the purpose."[1]

Admiral Moffett also pointed out that in the not too distant future "… commercial air-lines would be operating between most of the most important cities of the county. Communities that are preapred (sic) to establish connections with the least delay naturally would derive first advantages therefrom."[2] He also noted that with the government expanding airmail service "…it would not be surprising to receive a request at almost any time for landing facilities at Pensacola."[3] He was proved right on both accounts.

As the leases for Old Corry Field approached expiration on July 1, 1927, an editorial in *The Pensacola Journal* recommended that the city purchase the field. The editor noted that "The option on the Corry Field property, as is generally known, calss (sic) for its purchase at most reasonable figures."[4]

No action was taken and Old Corry Field remained in private hands. However, an editorial in March 1928 noted:

Pensacola has the Naval Air Station to add interest to aviation activities, and in the old Corry Field, owned by private interest, has a landing place. But if this city is to keep abreast in aviation, in which it should take leadership, some active movement to that end should be put under way. It took Jacksonville three years to get its airport but it is there. It may take Pensacola three years, but now is the time to start.[5]

ROAD TO RUIN OLD CORRY FIELD

With most of Old Corry Field lying fallow, the county announced plans in April 1928 to build a road through the property. Municipal airport advocates argued that the road would render the field "...practically valueless for airplane landings."[6] *The Pensacola Journal* disagreed and lambasted those opposed to this construction by stating in an editorial:

Old Corry field is private property. The city has no definite right to say what shall be done with it. The Journal believes it should be acquired by the city and made a municipal airport, but this has not been done yet and may never be done. Running a road through it will not prevent its use later as a landing field anyhow.[7]

The owner of one section of the property also protested to the country commissioners. M.C. Boley stated that "The new Ferry Pass road will cut Corry Field in half and make it unfit for a municipal airport if one is ever established."[8] However, the project engineer for the new road countered that "Enough land will remain to serve for commercial flying."[9]

One particularly perceptive letter to the editor noted:

1. If this situation will be carefully examined as has been done by the writer, it can be seen that on either side of the proposed location of the road there is sufficient room for a commercial flying field.

2. Is there any movement on foot for the establishment of a land flying field, if so, is it proposed to raise the money by popular subscription or by count bond issues; also, if so, has the owner of this land quoted a satisfactory price and will he enter into an agreement to sell this land for a commercial flying field at this price? It seems to me that this community not so very long ago had a serious disagreement with the owner of this land as to a satisfactory price.

3. If a commercial aviation field is to be established are we to believe that this particular site is the only one available? There is nearly one-half million acres of land in Escambia County, Florida, most of it undeveloped, and destined to remain so unless we can work together and avoid obstructing every forward movement.[10]

Mr. Pierce, author of this letter, had perfectly captured the issue. If the citizens of Pensacola truly wanted a municipal airport, then they should either buy the entire property of Old Corry Field, which had just been rejected by the U.S. Navy as unsuitable, or look elsewhere in the city for airport property. As it turns out, the latter course of action was selected only a few years later.[a]

U.S. NAVY SUPPORTS LANDING FIELDS OUTSIDE PENSACOLA

While Pensacola's civic leaders danced around the topic of a city-owned municipal airport, other towns and cities were not idle. As the reliability of the Naval aviator's aeroplanes increased, cross-country flights to practice

---

[a] Eventually, the road was built after the county took successful legal action to stop injunctions brought by one of the property owners.

navigational skills stretched further and further afield from Pensacola. To support this training, senior officers at Naval Air Station Pensacola envisioned an expanding network of outlying civilian landing fields available for use by their flight students.

One of the strongest advocates of this concept was Rear Admiral J.J. Raby.[b] During a Pensacola radio broadcast in December 1928 Admiral Raby "Commended the efforts of nearby towns to promote aviation by the establishment of airports...[and]...mentioned such accomplishments as municipal landing fields recently opened at Foley, Ala., Defuniak Springs, Mobile, Crestview and other points within a cruising radius of Pensacola."[11] In honor of Admiral Raby's efforts on their behalf, the city of Pascagoula, Mississippi dedicated their new airport as Raby Field.[12]

These airports served more important roles than just simple navigational homing points; they also offered a modest economic return to these communities. Naval, Army and civilian aviators visited these airfields and spent money for fuel and support. Later on, as commercial aviation grew, they also provided ready-made facilities for airmail and airline service.

Naval aviators from Pensacola were also in great demand for their experience and civil engineering skills in selecting landing field sites. *The Pensacola Journal* wrote:

Letters continue to come in to the air station from surrounding cities and towns regarding the opening of airports, the voting for bonds for an airport, the selection of sites for an airport or just the dream, hope or talk of an airport. Sure and steadily the various municipalities are getting air-minded and they are all realizing the benefits that will follow the establishing and equipping of an officially approved landing field.

The latest letter to be received at the air station is from R.E. Johnston, secretary of the chamber of commerce of Columbus, Miss. They are trying to arouse local interest in a proposed field and ask if such a field would be used by Pensacola planes on cross-county flights. They cite the case of Hattiesburg, Miss, which recently opened and has been visited on several occasion by planes from Pensacola.[13]

VISIT OF MARSHALL O. HOPPIN

In February 1929 Marshall O. Hoppin, a field worker of the airports division of the aeronautics branch of the Department of Commerce, visited Pensacola and re-ignited the issue of a commercial landing field for Pensacola.[c] *The Pensacola Journal* reported his visit:

---

[b] Rear Admiral Raby served two tours as commandant of NAS Pensacola and his son was also a Naval aviator. Of interest, there is a Raby Street off Barrancas Boulevard near the Naval Air Station.

[c] The Air Commerce Act of 1926 assigned federal aviation regulation to the Aeronautical Branch of the Department of Commerce.

Marshall O. Hoppin yesterday expressed surprise that Pensacola has no municipal airport, particularly in view of the city's familiarity with the importance of aviation. Hoppin was in Pensacola Thursday and Friday. He is a field worker for the bureau of aeronautics of the Department of Commerce. He is touring the south in the interest of establishing airports.

"Pensacola is ideally situated for a center city in air travel," Hoppin said. "What you should do here is make an airport for both land and seaplanes. Air travel isn't going to be confined to land planes. It will include seaplanes, and the city that can have water and land fields is a fortunate city indeed."

Hoppin said one air line which has been talked of in government circles would be from Miami, across to Tampa and up the fringe of the Gulf Coast through Pensacola to Mobile and New Orleans. At Mobile, it would connect with the already established New Orleans-Atlanta air mail. It also would link up at New Orleans with a line from Brownsville, Tex., which will be put into operation around the western fringe of the gulf. [14]

An additional comment by Hoppin that Pensacola is "Just like a lot of other Southern cities---sound asleep" when it came to establishing a municipal airport, prompted a quick response from O.H.L. Wernicke, chairman of the Greater Pensacola and District Movement.[15] Wernicke stated that Hoppin was unaware of his organization's plans to launch a $1,000,000 advertising campaign to capitalize on "...Pensacola's strategic position on the airmap of the United States."[16] He showed Hoppin a folder that depicted Pensacola's distance "as the crow flies" from various cities within the United States. Upon seeing this document, Hoppin "Pulled Pensacola's name from the list of 'sound asleep' Southern cities," and stated he was "...more enthusiastic over Pensacola's prospects in the world of the air."[17] Hoppin's impressions were considered important as he was thought to have considerable influence within the Department of Commerce.

Even Mobile chirped in when L.W. James, director of the Mobile office of the Bureau of Foreign and Domestic Commerce of the Department of Commerce, noted that "...Pensacola needs (a) commercial airport" and that he was "...particularly surprised that Pensacola has no commercial airport activity or facilities here."[18]

Just as the Kiwanis Club had stepped in years before to establish Pensacola's first commercial landing field, another organization was needed to step forward to establish a municipal airport. That organization turned out to be the Pensacola Lions Club.

PENSACOLA LIONS CLUB

"Pensacola needs a municipal airport in addition to the air travel facilities already offered here by the naval air station and Corry field" said George E. Hoffman, chairman of a special airport committee formed by the Pensacola Lions Club.[19] Hoppin had assured Pensacola the support of Col. S.H. Blee, chief of the airports division of the Department of Commerce, who would offer "...suggestions as to the most effective methods to follow in working for the establishment of an airport here during 1929."[20] The Pensacola Lions Club's Pensacola Municipal Airport project was for "...the use of heavier-than-air machines. It also includes the establishment of a mooring mast of lighter-than-air machines."[21] Their slogan was "An Airport for Pensacola in 1929."[22]

The mayor of Pensacola, Harvey Bayliss, suggested that the Lions Club "...arrange with some owner of land on the outskirts of the city to lease to the club for a nominal sum a tract of about 100 acres, to be used as a temporary municipal airport."[23] As had been done with both Corry Fields in the past, he indicated that the city would assume responsibility for such leases from the Lions Club.

"BUILD THE BIRD-HOUSES AND THE BIRDS WILL COME TO THEM"

However, the whole issue of a lack of airports in the south was larger than just Pensacola's shortcoming. Airmail, airways, airports, and airlines were multiplying in the northern part of the United States while the south seemed to be lagging behind significantly. Seeking to solve the mystery of what comes first, airports or airlines, an editorial in *The Pensacola Journal* in March 1929 noted:

The primary need of the south in aviation is better airports and more of them. Then there must be well-marked airways for day and night flying, and emergency, or inter-mediate fields where planes can come down and take off safely in case of forced landing, caused by motor trouble, bad weather, lack of fuel or other difficulty. When trouble occurs, a level, open space within gliding distance may save the plane and its occu-pants. "Build the bird-houses and the birds will come to them"; build the airports and the planes will use them. The south should share in the benefits and do its part in the development of the new system of transportation and the new commerce and industry represented by aviation. [24]

U.S. NAVY POLICY ON AIRPORT DEDICATIONS

The major highlight of any airport dedication was the appearance of aeroplanes from the Naval Air Station. So many requests were received from communities all over the south that eventually the U.S. Navy published a policy on their participation. Whether these regulations were designed to protect spectators from harm or to limit the enthusiasm of Navy pilots is hard to tell. Specifically, there would be "...*no wing-walking, airplane trapeze performing, changing from one plane to another or from one plane to some other vehicle or vice versa, delayed-release parachute jumping, or extremely hazardous 'stunt' maneuvers. Neither will Navy planes take part if admission is charged to spectators and private parties or concerns reap a profit.*"

PENSACOLA'S FIRST FLYING CLUB

The Pensacola Lions Club was aided in its efforts to establish Pensacola Municipal Airport by the formation of the city's first Aero Club in April 1929. Chartered under the American Society for the Promotion of Aviation, the new club published two goals.[25] The first goal was to "…cooperate with civic clubs of the city to …work together to obtain a municipal airport." The second goal was to "…obtain at least one plane, possibly from the government, for training purposes." Flight training was stated as "One of the main objectives of the clubs after equipment and an airport are established." [26]

PENSACOLA MUNICIPAL AIRPORT BILL

On May 6, 1929 the Pensacola Lions Club introduced a bill to the city commission "…which if passed by the state legislature, would give the city of Pensacola authority to provide a municipal airport."[27] The city commission passed the bill and it was forwarded to the Florida state legistature for approval. On May 14, 1929 the senate passed the bill that "…would authorize the city of Pensacola to issue bonds for the operation of a municipal airport."[28] On May 23, 1929 Governor Doyle E. Carlton signed the bill into law.[29] The bill authorized Pensacola "…to construct an airport if and when the city is financially able. The bill provides that the port does not have to be within the city limits."[30]

The Pensacola Lions were so thrilled they started making plans for an 'airport day' to stimulate "…interest in the plans of the Lions for a municipal air port."[31] However, Adrian E. Langford of the city commission stated that "Although bonds for $100,000 to buy and finance such a project have been authorized, we could not sell the bonds at this time, and if we did, it would only serve to further burden the taxpayers. This we are avoiding, and we believe the city will be well served by this plan."[32]

AVIATION SIGNS FOR PENSACOLA

If Pensacola didn't yet have a municipal airport of her own at least she could identify the city to aviators flying overhead. On September 5, 1929, *The Pensacola Journal* noted that "Pensacola's name will be painted in 15-foot letters on top of the grand stand at Legion field for the benefit of airplane pilots…(and)…an arrow will point to a landing field."[33] It is unclear exactly what landing field the arrow would point to, but it is assumed that it was Old Corry Field.

Ten days later, on September 15, 1929 *The Pensacola Journal* reported that the Standard Oil Company[d] planned to mark the "Jacksonville-Pensacola-New Orleans airways route…by lettering the roofs of its plants along the way."[34] Standard Oil plants in Florida at McClenny, Live Oak, Madison, Perry, and

---

[d] The Standard Oil Company's cooperation with the U.S. Army Air Service to mark roofs in California was so successful that it began to mark roofs at its facilities across the United States.

Pensacola would be marked as well as plants in Mississippi at Pascagoula, Biloxi, Gulfport and Bay St. Louis.

CITY FIELD

On September 19, 1929 an article appeared in *The Pensacola Journal* about discussions between the Pensacola Lions Club and Texas Air Transport Co. regarding potential airmail service between Pensacola and Jacksonville.[35] A civilian airport in Pensacola would be needed to support such operations,[e] but Mayor J.H. Bayliss was quoted as stating that "...no plans for the establishing of a municipal airport were under way. If such plans materialize a part of old Corry field will probably be leased."[36]

Only five days later, on September 24, 1929 *The Pensacola Journal* announced that "Work will be started this week on Pensacola's long sought municipal airport."[37] As predicted by Mayor Bayliss, the site selected was a 100-acre

Figure 2. Enacted on May 23, 1929, this bill authorized the City of Pensacola to issue bonds to operate an airport with an airplane landing field and lighter-than-air mooring masts. (Courtesy Forida State Archives).

section of Old Corry Field lying east of the Ferry Pass Road. The land was owned by Nettie F. Ripple and she charged the Chamber of Commerce one dollar a year for a five year lease.[38] *The Pensacola Journal* pointed out that Ripple "...lives in Oklahoma City. Apparently, distance doesn't dim her civic spirit. Her example could well be followed in other cases."[39]

The city prepared to build a small building at the field to keep a telephone and first aid supplies. The total investment for the five year lease was expected to be $100, including the placement of markers "Sufficiently large to be seen from the air" since the airport was to be indicated on all National Airways maps.[40]

Pensacola Municipal Airport was not quite the modern, expansive facility envisioned by its supporters, yet an editorial in *The Pensacola Journal* noted:

The building to be erected at the city's airport won't be anything to take pictures of to send the folks back home. In fact it will be a small frame structure, costing about $100, to hold a telephone, medical supplies, etc. But nevertheless it enables the city to be classed among those which have municipal airport facilities and the city will be put on the airways map. This is important and will be more important as time passes.

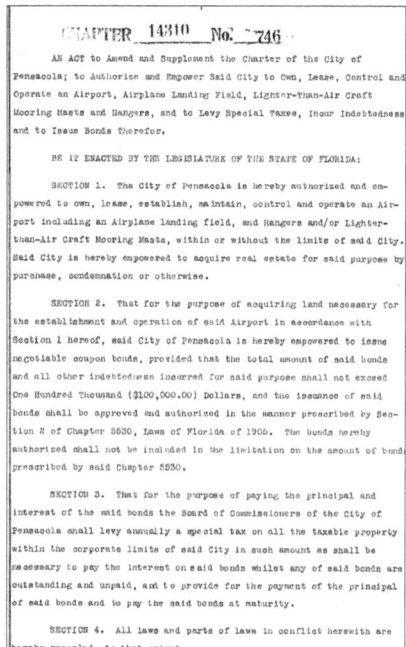

---

[e] While gracious in their occasional accommodation of civilian aeroplanes at their military facilities, the U.S. Navy could not support continuous civilian operations that would interfere with their primary mission of flight training.

PENSACOLA REGIONAL AIRPORT

The modern, well-equipped Pensacola Regional Airport we see today was born at the height of the Great Depression and built by the hands of relief workers. In November 1933, Pensacola officials opened negotiations with Francis W. Taylor to lease 500 acres of his property located northeast of the city as a municipal airport. Taylor, an early aviation visionary, had acquired "the old Baar property" after surveying the area for the most suitable land to establish an airport. A five-year lease was agreed upon at $1 a year, the U.S. Navy provided a discarded hangar valued at $1,500, and the county consented to build a connecting road from 12th Avenue. Funds from the Civil Work Administration (CWA), a short-term emergency relief program created to help the unemployed, were obtained to pay several hundred men to clear and grade the property. Actual work began at the site in December 1933 and by January 1934 about half of the 500-acre site had been prepared.

CWA projects were eventually assumed by the Works Progress Administration (WPA). Under the WPA public funds could not be spent on improvements to property the city did not own and all worked ceased at the airport. Florida Senator Phillip D. Beall, in an appearance in Pensacola in August 1935, urged the city to purchase the property in order to continue government support for construction at the field.

Negotiations for the purchase of the property began in late 1935 and assumed a greater importance when the U.S. Navy stated that due to increased training requirements they needed another landing field immediately or they might take their training elsewhere. After rigorous debate, on October 28, 1935 the city council voted to buy the airport for $50,000 from Taylor. Almost immediately, the WPA authorized $129,350 for improvements to the airport, including the construction of an airport administration building and three paved runways. In January 1936 the airport was leased to the U.S. Navy, with the understanding that the city reserved the right to use the field as a municipal airport.

In March 1936 a request by C.H. (Harry) Blanchard of Blanchard Air Service to assume management of the airport was approved and an agreement was reached on revenue sharing. Blanchard served as airport manager until 1962, with the exception of a short period during World War II when he was overseas in the Army Air Service. During this period his wife, Van, served as acting airport manager.

For several decades the airport was known as Hagler Field in honor of L.C. Hagler who was mayor of Pensacola and manager of the San Carlos Hotel. It was not until fairly recently that the field was renamed Pensacola Regional Airport.

Planes of all types, commercial and transient, can land at the municipal airport which will be shown on the maps that guide the nation's air travel. Old Corry Field, while thus rated as Pensacola's municipal airport, serves a double purpose with naval fliers using it as a training field.

The Chamber of Commerce and city officials who have thus provided an arrangement that puts Pensacola on the airways maps with other progressive cities.

The time will come when Pensacola will have a big municipal airport with hangars and all equipment. That, however, will cost much money and may be far in the distance. In the meantime, the city is well off to have an airport rating with virtually no cost.[41]

Although listed as Pensacola Municipal Airport, the airfield was known by local aviators as "City Field"

THE FUTURE

The Pensacola Lions Club had settled for a temporary municipal airport at Old Corry Field but they still very much wanted to "Secure for Pensacola a permanent airport."[42] An editorial in *The Pensacola Journal* noted:

Pensacola today has one of the largest naval air stations in the world, with a $5,000,000 program of expansion now taking form, but this city needs something more than a naval airport.

Recognizing the importance of commercial aviation and the strides being made in this direction, the Lions club is seeking to establish a municipal airport, steps towards which the city has already taken, with the leasing of old Corry Field.

The navy is sharing this field, as it properly should, and it is fine thing to have gone as far in connection with the field as has been made possible.

But the Lions club is desirous of seeing a municipal airport at Pensacola ranking with those of other cities, and to that end is urging all other civic clubs to join in this movement.

The plan of the Lions club is to establish a permanent Inter-Club Airport committee, believing that much remains to be accomplished, which such a committee might help to bring about.

Even the small towns of the country are now establishing such airports and these are not only desirable form a civic standpoint, but in time will become necessary to any city of size. Commercial aviation is making remarkable progress, and the time will come when Pensacola not only will have an airport commensurate to such needs, but will be a center for manufacture of planes.

Such an organization as is suggested by the Lions club can do much to bring about both the airport and the factory.[43]

Little did the Pensacola Lions Club realize just how long it would take before tiny City Field would be replaced by a modern, expansive Pensacola Municipal Airport, situated at an entirely new location east of their field.

NOTES

[1] Pensacola Soon Will Need Commercial Plane Field. (1927, May 17). *The Pensacola Journal.*

[2] Pensacola Soon Will Need Commercial Plane Field. (1927, May 17). *The Pensacola Journal.*

[3] Pensacola Soon Will Need Commercial Plane Field. (1927, May 17). *The Pensacola Journal.*

[4] Pensacola Soon Will Need Commercial Plane Field. (1927, May 17). *The Pensacola Journal.*

[5] Jacksonville Takes The Air. (1928, February 12). *The Pensacola Journal.*

[6] Fear Road May Ruin Field For Use By Planes. (1928, April 4). *The Pensacola Journal.*

[7] More Needless Delay. (1928, April 18). *The Pensacola Journal.*

[8] Owner Speaks For Air Field. (1928, April 18). *The Pensacola Journal.*

[9] Owner Speaks For Air Field. (1928, April 18). *The Pensacola Journal.*

[10] Pierce, P.L. (1928, April 19). Ferry Pass and Flying Field. *The Pensacola Journal.*

[11] Admiral Looks For Expansion Of Air Station. (1928, December 20). *The Pensacola Journal.*

[12] Airplane Field At Pascagoula Named For Raby. (1925, November 13). *The Pensacola Journal.*

[13] Barber, B.B. (1929, June 2). Daily News Of The Air Station. *The Pensacola Journal.*

[14] Man Says City Surprises Him. (1929, February 16). *The Pensacola Journal.*

[15] Pensacola Has Great Future As An Airport. (1929, February 17). *The Pensacola Journal.*

[16] Pensacola Has Great Future As An Airport. (1929, February 17). *The Pensacola Journal.*

[17] Pensacola Has Great Future As An Airport. (1929, February 17). *The Pensacola Journal.*

[18] Pensacola Has One Big Fault. (1929, February 24). *The Pensacola Journal.*

[19] Lions To Boost Airport Plans In Pensacola. (1929, February 24). *The Pensacola Journal.*

[20] Lions To Boost Airport Plans In Pensacola. (1929, February 24). *The Pensacola Journal.*

[21] Lions To Boost Airport Plans In Pensacola. (1929, February 24). *The Pensacola Journal.*

[22] City Approves Airport Bill. (1929, May 7). *The Pensacola Journal.*

[23] Bayliss Pats Lions' Heads For Air Idea. (1929, March 2). *The Pensacola Journal.*

[24] The Primary Need. (1929, March 21). *The Pensacola Journal.*

[25] Pensacola Has New Aero Club. (1929, May 5). *The Pensacola Journal.*

[26] Pensacola Has New Aero Club. (1929, May 5). *The Pensacola Journal.*

[27] City Approves Airport Bill. (1929, May 7). *The Pensacola Journal.*

[28] Senate Passes Airport Bill. (1929, May 15). *The Pensacola Journal.*

[29] Lion's Airport Bill Is Law. (1929, May 28). *The Pensacola Journal.*

[30] Lion's Airport Bill Is Law. (1929, May 28). *The Pensacola Journal.*

[31] Lions Prepare Airport Boost. (1929, May 29). *The Pensacola Journal.*

[32] Airport Work Starts During Present Week. (1929, September 24). *The Pensacola Journal.*

[33] City's Name Can Be Read From Air. (1929, September 5). *The Pensacola Journal.*

[34] Pensacola To Have Airway Markings. (1929, September 15). *The Pensacola Journal.*

[35] Lions Plan On Air Mail Route For Pensacola. (1929, September 19). *The Pensacola Journal.*

[36] Lions Plan On Air Mail Route For Pensacola. (1929, September 19). *The Pensacola Journal.*

[37] Airport Work Starts During Present Week. (1929, September 24). *The Pensacola Journal.*

[38] Pensacola To Have Municipal Airport. (1929, September 20). *Air Station News, Pensacola, Florida.*

[39] As It Seems To Us. (1929, September 21). *The Pensacola Journal.*

[40] Airport Work Starts During Present Week. (1929, September 24). *The Pensacola Journal.*

[41] The City Airport. (1929, September 25). *The Pensacola Journal.*

[42] Civic Leaders Will Meet To Talk Aviation. (1929, October 11). *The Pensacola Journal.*

[43] An Inter-club Movement. (1929, October 16). *The Pensacola Journal.*

CHAPTER 19
AIRMAIL

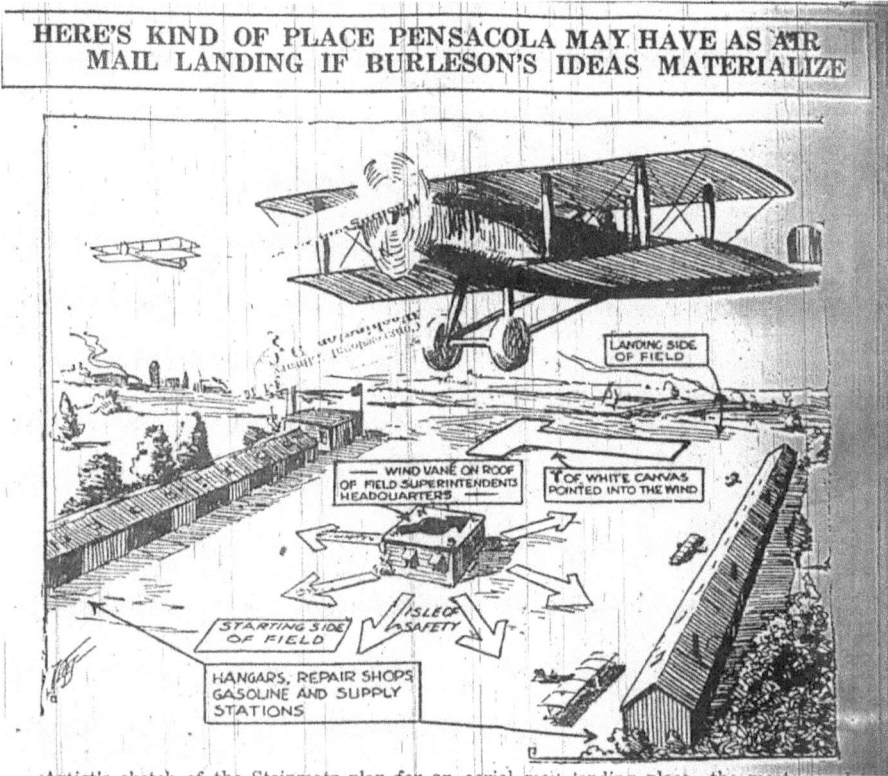

Figure 1. November 1918 drawing in *The Pensacola Journal* depicting a possible airmail landing field for Pensacola. (Courtesy *Pensacola News Journal*)

FIRST OFFICIAL AIRMAIL IN THE UNITED STATES

The first official airmail in the United States was flown as part of a publicity stunt during a nine day aviation meet at Garden City, Long Island on September 23, 1911.[1] The Postmaster General gave official permission to set up a temporary post office branch in tents at the airfield and mail was transported in pouches aboard a Bleriot monoplane piloted by Earle Ovington. Ovington flew three miles to Mineola, Long Island, where he dropped the pouch in a field for retrieval by the local postmaster, who entered it into the regular postal system. He would then return to Garden City for more mail and further trips.[2]

In 1911, this three-mile trip pretty much represented the technical limit of the aeroplanes of the time. In 1834 Postmaster General William T. Barry said that "The celerity of the mail should always be equal to the most rapid transition of the traveler."[3] While the speed of aeroplanes may have appeared attractive, their limited range, small cargo capability, and frequent breakdowns could not be substituted for the steadfast reliability of trains.

AIRMAIL SERVICE INITIATED

By 1918, airmail proponents were encouraged by the improved performance of aeroplanes developed during World War I. Bolstered by Congressional appropriations, they convinced the Post Office and War Department to cooperate in a joint experimental airmail route between New York, Philadelphia, and Washington, D.C. The 204-mile trip was estimated to take three hours. The U.S. Army provided planes, pilots, and the Washington landing field, while the Post Office provided fuel, New York and Philadelphia landing fields and staff.

On May 15, 1918 witnessed by a host of dignitaries including President Woodrow Wilson, a Curtiss JN-4H piloted by Second Lieutenant George Boyle, U.S. Army, lifted off from Washington, D.C. in the United States' first regular airmail service.[4] For the next several months, the Post Office would use U.S. Army pilots and aeroplanes to operate its airmail service.

On August 12, 1918, the Post Office Department assumed responsibility for the delivery of airmail, contracting for its own aeroplanes and civilian pilots. On that day, Max Miller, the first pilot hired by the Post Office, piloted a Curtiss R-4 from College Park, Maryland to New York to make the first flight operated by the Post Office.

From this simple route, airmail expanded rapidly to dozens of cities in the northeast. However, airmail service for the south lagged far behind, mostly due to a lack of suitable airfields. Even transcontinental airmail routes were pioneered before airmail finally worked its way down the eastern seaboard to Florida.[a]

Figure 2. The first airmail stamp issued was the 24 cent red, white and blue Jenny, representing a per ounce cost of 14 cents for an air fee and 10 cents for a special delivery fee. The green and white 16-cent stamp Jenny was issued a few months later when the air fee dropped to 6 cents and the special delivery fee remained at 10 cents. A few months later the 6 cent stamp was issued, reflecting the complete drop of the special delivery fee, making the total airmail cost per ounce in 1918 6 cents. (Courtesy of author)

---

[a] The first transcontinental airmail flight was made in February 1921. The first contract airmail route in Florida was not flown until April 1, 1926.

AIRMAIL ROUTES

In late 1918 the United States Post Office started planning for their vision of post-World War I airmail routes that would service the entire United States and South America. Once aeroplanes and aviation equipment were released from service, four major airmail trunk lines would be established: New York to San Francisco, Boston to Key West, Key West to Panama, and Key West to South America.[5] Individual cities designated as stopovers would be announced later. While Pensacola might not hope to be a station on an airmail route, she could certainly hope to be over flown by one, with the potential for added service sometime in the future.

This was not to be. The nearest announced air route stop to Pensacola was Birmingham, Alabama.[6] Pensacola was not located on any of the first planned airmail routes, nor was she located on any planned feeder lines designed to expand airmail service to additional cities.

AIRMAIL FOR PENSACOLA

Let there be no doubt that Pensacola desperately wanted airmail service. On May 14, 1918 a day before the first regular airmail service in the United States was inaugurated, *The Pensacola Journal* said that the local post office would "...likely be overrun with demands of patrons for new aeraplane (sic) postage stamps."[7] The article also noted that if the airmail service "...proved successful, the likelihood is Pensacola will be included in any additional services established."[8]

Less than a month later, another newspaper article appeared noting that authorities in Washington, D.C. had been contacted regarding "...the establishment of an airmail service from Pensacola to Mobile, Biloxi, Gulfport and New Orleans, all of these cities being points where there are ship-building plants."[9]

In November 1918, Pensacola felt certain that she would be included in plans as a "Landing place for the U.S. aerial mail service, to be established between Florida and northern points."[10] Exactly when this event might occur was a subject of serious speculation by Pensacolians.

Pensacola's own postmasters were tireless in their attempts to obtain airmail service for Pensacola. In 1918, Pensacola Postmaster B.S. Hancock met with both the chief clerk of the railway mail service[11] and representatives of the Aerial League of America[12] to advance Pensacola's opportunities for airmail service.

All of these initial actions appeared promising but the same encouraging developments regarding airmail service for Pensacola would be reported in *The Pensacola Journal* for almost twenty years. Time after time Pensacolians would have their hopes for airmail service raised only to find them dashed by the cold reality of economics. Although ultimately successful, the future for Pensacola's airmail advocates would be filled with disappointments, false starts and empty promises.

AIRMAIL COST			
Year	Service	Cost	2002 Equivalent
1918	Airmail	.24	2.86
1918	Airmail	.16	1.90
1918	Airmail	.06	.71
1918	Regular	.03	.41
1928	Airmail	.05	.52
1928	Regular	.02	.21

When a comparison is made between the cost of an airmail stamp and a regular domestic stamp in 1918 it is easy to see why the U.S. Post Office experienced problems generating an adequate volume of airmail.

After the initial 24-cent stamp was issued in 1918, the Post Office made several rate changes to lower the price of an airmail stamp. By 1928, for example, the airmail rate was only 5 cents but this was still twice the cost of regular mail.

Shown above is the "Inverted Jenny." One hundred of these stamps were accidentally issued before the printing error was noticed. Their value today is measured in millions of dollars.

AIRMAIL SERVICE PRIVATIZED

Pensacola's difficulty in establishing airmail service can be traced to the passage of the Airmail Act of 1925. More commonly known as the Contract Airmail Act, this legislation sought to encourage the development of commercial aviation by authorizing the Post Office to transfer airmail contracts to private hands. Any company or individual could bid on contract airmail routes, commonly abbreviated as CAMs. Governed by various acts of Congress, the decision to award contract airmail routes rested solely with the second assistant postmaster general.[b] For the most part, they were awarded to the lowest bidder, subject to specific contract performance requirements.

Airmail in the late 1920s was a risky business. Fields were primitive, navigation aids almost nonexistent, and the risk of inadequate revenue a constant worry. The successful contractor assumed responsibility for landing fields, emergency landing fields, route and city selection, aids to navigation, communication, and aircraft. The Post Office's only responsibility was the transport of the mail to and from the flying fields and weighing it.

Which contract airmail route to bid on was left solely to interested parties. Prospective contractors were looking for those routes that presented a

[b] One of several assistants to the postmaster general, the second assistant postmaster general's responsibilities are the transportation of mail both by air and surface.

potential for carrying a sufficient volume of airmail to generate a profit. Consequently, smaller towns were bypassed in favor of the larger cities.

In 1927, W. Irving Glover, Second Assistant Postmaster General, responding to a letter from F. M. Blount of the Pensacola Chamber of Commerce, clarified the United States Post Office's position concerning airmail service:

The policy of the Department in this respect is to authorize airmail service between any two points where the same may be found feasible. In this work we find that the situation is somewhat analogous to the old question of the railroad. The Department does not set up the equipment or operating facilities, but simply contracts with any individual, firm or corporation for the transportation of airmail. Obviously there are certain routes which appeal to aviation companies, particularly those between the larger cities where the saving of time is so great as to attract sufficient traffic to make the line profitable. Routes such as these have been established wherever the Department is aware there is some opportunity for a contractor to receive enough airmail to successfully operate the service.

The office has not fixed any certain amount of airmail that must be available before service is established. The amount of airmail necessary in order to warrant the establishment of a route depends entirely upon the length of the proposed route. [13]

Therefore, before any air carrier would initiate a bid to provide airmail service, cities along the route would have to demonstrate that the volume of airmail developed would be sufficient to generate a profit. Several airmail companies went bankrupt because the promised volume of airmail was never fulfilled.

Further, for the individual citizen, the cost of airmail was expensive when compared to regular mail rates. For example, the cost of an airmail letter in 1928 was 5 cents per ounce, over twice the price of a normal domestic rate of 2 cents per ounce.

PENSACOLA AIRMAIL

Nonetheless, Pensacola just could not understand why she was not included in any airmail routes. In 1925 *The Pensacola Journal* pointed out:

The location in Pensacola of the largest naval airplane training station in the country and of the army forces, together with shipping and other commercial activities, is expected to be strongly considered by post office officials in connection with the proposed airmail service from the Gulf to the north and east. Two large landplane fields are located here, Corry Field and the field at the Naval Air Station.[14]

Other communities approached Pensacola to cooperate with establishing airmail routes but none would see fruition. In one example, Birmingham approached Pensacola Postmaster Milton E. Clark in May 1925 to have Pensacola "...become the southern terminus of a proposed airmail service, to New York City by way of Birmingham, Nashville, Louisville and Cleveland."[15]

However, in this "for profit" airmail environment, Pensacola's major problem, besides the lack of a good commercial landing field, was the low potential for an adequate volume of airmail. While this would prevent direct service from the southern terminuses of any of the air carriers, Pensacola could certainly hope to be added as an enroute stop of any airmail service flown nearby. Surprisingly, the nearest airmail stop to Pensacola turned out to be Mobile, Alabama, who received service on May 1, 1928 even though airmail had arrived in the state of Florida a few years earlier.

FIRST AIRMAIL SERVICE IN FLORIDA

Figure 3. A Travel Air 2000 of Florida Airways at Jacksonville in 1926. This airline carried the first airmail in Florida.(Courtesy Florida State Archives)

On February 11, 1926 the first contract airmail route in Florida, CAM-10, was awarded to Florida Airways Corporation.[16] To be flown between Atlanta and Miami via Macon, Jacksonville, Tampa, and Fort Myers, with not less than six round trips weekly, Florida Airways was to receive $3 per pound of airmail.[17]

Almost immediately Florida Airways Corporation announced:

Extensive operation of air service over Florida, and into other states is being planned, the president of the corporation said. Not only has the government indicated that it would award the corporation the mail contract, but routes to include Havana, Cuba,

Figure 4. The first airmail to Florida was flown by Florida Airways on CAM-10 from Atlanta to Jacksonville on September 15, 1926. (Courtesy of author)

to Dallas, Texas, Jacksonville, Tallahassee, Pensacola and New Orleans and Havana to Memphis by way of Atlanta are being worked out.[18]

The future looked promising for airmail to Pensacola. Florida Airways' first airmail flight between Jacksonville and Tampa launched on April 1, 1926 and service was established with Atlanta on September 15, 1926.

Unfortunately, Florida Airways' airmail route did not connect with any northbound or westbound routes out of Atlanta or any southbound routes out of Miami. By late 1926, Florida Airways stopped flying airmail altogether, a victim of low airmail volume, and the Post Office cancelled its contract. Florida Airways was insolvent by the end of the year.

Airmail service was not resumed in Florida until Pitcairn Aviation[c] acquired Florida Airways' abandoned CAM-10. Renumbered CAM-25, service began on December 1, 1928 from Atlanta to Miami.[19]

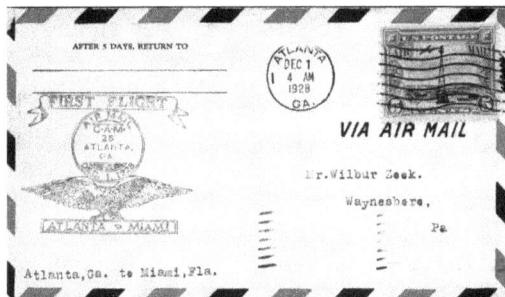

Figure 5. After the collapse of Florida Airways, the next airmail in Florida did not occur until Pitcairn Aviation was awarded the CAM-25 airmail route in late 1928. (Author's collection)

Development of airmail routes in Florida continued to progress slowly. Even as late as December 1929, there were only six cities in the state receiving airmail, all from Pitcairn Aviation: Jacksonville, Miami, Key West, Tampa, St. Petersburg and Orlando.[20]

PENSACOLA'S REQUIRED AIRMAIL VOLUME

In July 1927, Pensacola was advised that she might be included in a proposed airmail route that would fly from New Orleans to Atlanta by way of Birmingham, Mobile and Pensacola.[21] The huge caveat was that the amount of mail into and out of Pensacola had to justify the service. A strong point also was made that "Pensacola will not be included in the route immediately on its establishment but will later be admitted if the traffic here is of sufficient volume."[22]

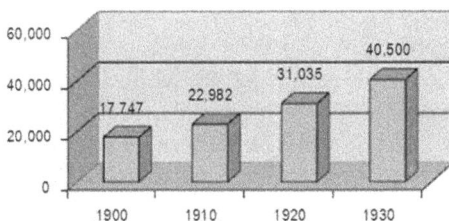

Pensacola's Population

Figure 6. Although Pensacola demonstrated a steady increase in population through 1930, there was not quite enough airmail volume to justify service for the city. (Courtesy of author)

The Pensacola Journal noted that:

Inclusion of Pensacola in the airmail route would greatly expedite the handling of letters and parcels between this city and New York City. A plane leaving New Orleans at 2 o'clock in the afternoon would be scheduled to arrive in New York City at 5 o'clock on the following morning.[23]

This proposal was added additional credibility when Pensacola was notified that a "United States mail airplane would arrive here next Wednesday on an unofficial experimental flight from New Orleans to Atlanta."[24] The goal was to generate as much publicity regarding airmail service as possible. The pilot of the experimental mail plane was Major Sumpter Smith "…well-known here, having been stationed here last summer as commander of the air division of

---

[c] Later to become Eastern Air Lines.

the Alabama national guard that spent two weeks in training at [old] Corry Field."[25] Major Smith landed at Old Corry Field shortly before 1 p.m. and was greeted by a large crowd and a multitude of local dignitaries.[26]

Hopes in Pensacola were raised to a fevered pitch a few days later when *The Pensacola Journal* reported that advertisements had been placed for airmail service along the route pioneered by Major Smith.[27] The exact amount of mail that would justify Pensacola's inclusion in this airmail service was detailed in *The Pensacola Journal*: "Pensacola will be expected to furnish five pounds[d] of mail per day."[28] *The Pensacola Journal* noted that "In the event the new mail line is routed through Pensacola a letter mailed here at 5 o'clock in the afternoon will be received in New York at 6:15 o'clock in the morning and will be in Boston at 5 o'clock."[29]

In an attempt to determine exactly how much mail Pensacola could possibly generate, the Chamber of Commerce undertook to survey all towns between Pensacola and Tallahassee. The reasoning was "…if the towns of the section contribute to the quota of the city, the desired results can be obtained."[30] It also was noted that "In case the city can not furnish this amount it will not be profitable for the planes to stop here."[31]

In their survey letter mailed to West Florida and South Alabama towns adjacent to Pensacola, the amount of necessary airmail was upped to a guarantee of seven pounds[e] of airmail daily.[32] However, the response must have been under-whelming as a later editorial in *The Pensacola Journal* noted that "Letters were sent out by the Chamber of Commerce and a tentative program was considered but nothing tangible resulted."[33]

CAM-23 OMITS PENSACOLA

In 1928 CAM-23 was approved between Atlanta and New Orleans with service scheduled to start on April 15, 1928. To be flown by St. Tammany Gulf Coast Airway, it included a stop in Mobile.[34] Pensacola was bypassed.

This blow to the egos of Pensacolians must have been shattering.

 A frustrated editorial noted that:

…Pensacola is making no effort to place itself on the airmail route. Which seems a pity, in view of the fact that this city is so well situated to take advantage of the opportunity offered in the establishment of the New Orleans-Mobile-Atlanta line.[35]

While the most likely reason for Pensacola's omission was a perceived lack of airmail volume, the reason pinpointed by an editorial was the lack of a commercial landing field. The newspaper noted that "It may be that Pensacola, with naval flying activities a part of its daily life, does not feel the need of an

---

[d] At an average letter weight of one half ounce, that would work out to 160 airmail letters per day, every day.

[e] Yipes! Now we're up to 224 letters per day.

airport as do many other cities."[36]

Nonetheless, it was announced that Pensacolians could access their first airmail using CAM-23.[37] The only hitch was that Pensacola's airmail had to be aboard the early morning train to Mobile in order to meet the connecting airmail aeroplane there. Hence, "Only letters mailed in the early morning before 8:30 a.m. would be hastened by the airmail service."[38] Since airmail was picked up in Mobile by a northbound aeroplane at 2:35 p.m., any mail posted later would not arrive in Mobile before the aeroplane departed. [39]

A way to avoid the restrictions of this train schedule was proposed by Pensacola Postmaster M.E. Clark. He suggested improving the roads between Mobile and Pensacola[f] and noted that "…with hard surfaced roads between Pensacola and Mobile…a motorcycle could make a quick trip and greatly expedite mail service…"[40]

Figure 7.   All airmail sent from Mobile on the inaugural flight of CAM-23 carried this special emblem. (Courtesy *The Mobile Register 1928©*. All rights reserved. Reprinted with permission).

Nonetheless, Pensacola was still thrilled to have achieved this small modicum of airmail success and the Postmaster even obtained an airmail postmark. On April 20, 1928 *The Pensacola Journal* heralded:

PENSACOLA NOW USES AIRMAILS

Postoffice Gets New Mark For Outgoing Post

Beginning yesterday morning all mail outward bound from Pensacola is postmarked with a new die which designates that this city's mail service is on an airmail line.

The new print is of an aeroplane with the reading 'air-mail saves time.' It was obtained by the post office at the request of Postmaster M.E. Clark and the Chamber of Commerce.

A savings of about 24 hours in the mail service between New York and Pensacola results from the re-cently announced New York-New Orleans line which goes into effect May 1.

Figure 8. Example of the postmark placed on Pensacola mail in 1928. This particular letter was postmarked in 1927. Note the use of the regular 2 cent stamp. An airmail stamp would have cost 5 cents. (Author's collection)

Stops will be made at Mobile where Pensacola mail will be packed up and dis-

[f] The road between Mobile and Pensacola would not be paved for several years.

charged.[41]

However, the people of Pensacola were not using this airmail opportunity as much as was anticipated. *The Pensacola Journal* noted:

WRITERS DON'T USE AIRMAILS

Postmaster Draws Parallel With New Telephone

R. A. Baker, assistant postmaster of the Pensacola post office yesterday compared the public's reluctance to patronize airmail service with its delay in turning the invention of the telephone to immediate profit. 'When the telephone was first invented it was spoken of as a rather expensive toy and was generally considered as being worth nothing more than that,' he said.

'It is the same with the airmail and I feel sure that mail service by air will continue to grow and it will be patronized more by the public.'

Speaking of the time saved to Pensacolians by the New York-New Orleans route, Captain Baker called attention to a letter received last week mailed in New York Friday at 7 p.m. and which arrived here 5 p.m. Saturday.[42]

FIRST AIRMAIL FLOWN FROM PENSACOLA

At this point, no airmail had actually flown from Pensacola, it simply rode the train to Mobile to be placed aboard an airmail plane. But in March 1929 rain waters drenched West Florida and Southern Alabama in the "worst flood in history." Pensacola was cut off from all rail and highway transportation. U.S. Navy planes assisted in surveying the extent of the flood and dropping supplies. They also carried first class mail to Mobile.[43] *The Pensacola Journal* reported:

CITY'S LETTERS RIDE IN PLANES AS WATERS RISE

Fliers Serve As Mailmen In Flood Crisis – All Mail By Mobile

Mail from Pensacola to other parts of the country is now going out by airplane. Yesterday one plane carried mail to New Orleans and another to Mobile.

All mail into Pensacola will be routed through Mobile until train operations are resumed, Poster M.E. Clark said last night

Thanks Childs

'It is only through the wonderful cooperation of Commander Warren G. Childs, acting commandant of the Naval Air station, that Pensacolians are able to communicate by letters with the rest of the world,' Postmaster Clark said.

He has agreed to furnish us with one plane trip each day for carrying mail to and from Mobile until regular communications with the rest of the country are re-established.

The mail plane provided by the air station will leave Pensacola this morning at 9:00 a.m. for Mobile.

Mail is Heavy

Figure 9. First documented airmail to fly from Pensacola occurred when the U.S. Navy provided seaplane service to Mobile after the floods of 1929 prevented access to Pensacola. (Author's collection)

The first plane which left here yesterday, on the way to New Orleans, carried 1,000 pounds of first class mail, Postmaster Clark said. The second plane, which left later for Mobile, carried 500 and 600 pounds of first class mail (sic).

Mail addressed to Pensacola is being routed through Birmingham to Mobile, Postmaster Clark said. From Mobile it will be brought here by the navy fliers.[44]

Pensacola Postmaster M.E. Clark, ever the optimist about establishing Pensacola's own contract airmail service, noted that:

Pensacola is getting the habit of mailing by airplane even more strongly since the floods sent all first class mail through air in navy planes…the volume of airmail out of here…has shown more rapid increases.[45]

He also noted:

Uncle Sam is doing everything he can to make the airmails popular. He has taken the lick out of the airmail stamps by now having them printed on the airmail envelopes. He has reduced the rate from 10 cents to five cents for the first ounce of each airmail communication.[46]

However, the postmaster also pointed out that if Pensacola wanted airmail they "…would have to send out an average of 300 airmail letters every day."[47] Since on September 19, 1929 Postmaster Clark stated that the current airmail volume out of Pensacola was about 25 airmail letters a day, it does not seem realistic to expect that a city of less than 40,000 residents would be able to generate that much airmail.

TEXAS AIR TRANSPORT COMPANY

Pensacola's next adventure with airmail occurred in September 1929. The Pensacola Lions Club, working hard to establish Pensacola Municipal Airport, announced that they had:

...received a letter from officials of the Texas Company stating that an engineer of the company would be sent here for the purpose of making a survey of the territory between here and Jacksonville. If the report of the engineer is favorable the company would lend every assistance to the city in establishing an air line between the two points. If facilities were provided, they would contract with the government to carry mail...[48]

What specific facilities were needed by Texas Air Transport[g] in Pensacola were not stated but certainly at the top of any list would be a commercial airport. As discussed previously, in September 1929 a small section of Old Corry Field was leased to establish Pensacola Municipal Airport. Unfortunately, Texas Air Transport decided to bypass Florida entirely and concentrate its operations in Texas.

THE FUTURE

Pensacolians would just have to be patient. Very patient, in fact. It was not until November 1, 1938 that National Airlines landed at the new Pensacola Municipal Airport located near Scenic Highway for Pensacola's first regularly scheduled airmail service.

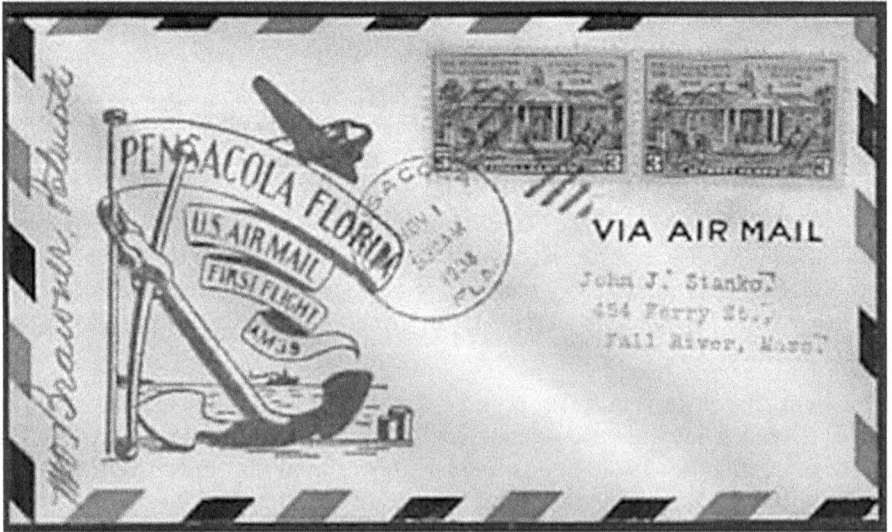

Figure 10. On November 1, 1938 Pensacola finally received her first daily airmail service courtesy of National Airlines. (Author's collection)

NOTES
[g] Eventually merged into the present American Airlines.

[1] Jackson, D.D. (1982) *Flying the Airmail.* Alexandria, VA: Time-Life Books, page 19.

[2] Jackson, D.D. (1982) Flying *the Airmail.* Alexandria, VA: Time-Life Books, page 19.

[3] Airmail's First Day. (1994). Retrieved from http://preview.thehistorynet.com/ aviationhistory/articles/0594_coverhtm.

[4] Jackson, D.D. (1982) *Flying the Airmail.* Alexandria, VA: Time-Life Books, page 25.

[5] Extension Of Air Service To Pensacola. (1918, December 6). *The Pensacola Journal.*

[6] Governor Catts Endorses Air-Mail For Florida. (1918, November 30). *The Pensacola Journal.*

[7] Aeroplane Stamps Be In Much Demand. (1918, May 14). *The Pensacola Journal.*

[8] Aeroplane Stamps Be In Much Demand. (1918, May 14). *The Pensacola Journal.*

[9] Airmail Service Recommended For Pensacola. (1918, June 4). *The Pensacola Journal.*

[10] Pensacola To Be Station On Aerial Mail. (1918, November 30). *The Pensacola Journal.*

[11] Airmail Service Recommended For Pensacola. (1918, June 4). *The Pensacola Journal.*

[12] Pensacola To Be Station On Aerial Mail. (1918, November 30). *The Pensacola Journal.*

[13] Glover, W. I. (Personal Communications, 1927, February 15). Special Collections. John C. Pace Library.

[14] Pensacola To Seek Aerial Mail. (1925, May 15). *The Pensacola Journal.*

[15] Request For Aid Received. (1925, May 13). *The Pensacola Journal.*

[16] Lazarus, W.C. (1942). *Wings in the Sun.* Orlando, FL: Tyn Cobb's Florida Press.

[17] Open Bids of Florida Airmail Service. (1926, January 19). *The Pensacola Journal.*

[18] Pensacola To Get Air Line Says Report. (1926, February 10). *The Pensacola Journal.*

[19] Lazarus, W.C. (1942). *Wings in the Sun.* Orlando, FL: Tyn Cobb's Florida Press, page 112.

[20] Lazarus, W.C. (1951). *Wings In The Sun.* Orlando, FL: Tyn Cobb's Florida Press, page 113.

[21] Pensacola May Be Station On Proposed Airmail Line. (1927, July 16). *The Pensacola Journal.*

[22] Pensacola May Be Station On Proposed Airmail Line. (1927, July 16). *The Pensacola Journal.*

[23] Pensacola May Be Station On Proposed Airmail Line. (1927, July 16). *The Pensacola Journal.*

[24] Mail Service Sending Plane In Experiment. (1927, July 24). *The Pensacola Journal.*

[25] Mail Plane To Arrive Today On Trial Trip. (1927, July 27). *The Pensacola Journal.*

[26] First Flight For Air Mail Is Made Here. (1927, July 28). *The Pensacola Journal.*

[27] Bids Asked On Airmail Route. (1927, August 12). *The Pensacola Journal.*

[28] Mayor To Confer Today With Seaburg On Mails. (1927, August 19). *The Pensacola Journal.*

[29] Mayor To Confer Today With Seaburg On Mails. (1927, August 19). *The Pensacola Journal.*

[30] West Florida Towns To Be Canvassed As To Amount Of Airmail To Be Furnished. (1927, August 20). *The Pensacola Journal.*

[31] West Florida Towns To Be Canvassed As To Amount Of Airmail To Be Furnished. (1927, August 20). *The Pensacola Journal.*

[32] Pensacola Is Given Chance For Airmail. (1927, September 13). *The Pensacola Journal.*

[33] Landing Fields A Necessity. (1928, March 20). *The Pensacola Journal.*

[34] Air Mail and Candler Field. Retrieved October 19, 2002 from http://ngeorgia.com/feature/ airmail.html

[35] Landing Fields A Necessity. (1928, March 20). *The Pensacola Journal.*

[36] Landing Fields A Necessity. (1928, March 20). *The Pensacola Journal.*

[37] Pensacola Can Have Airmail To North Now. (1928, April 7). *The Pensacola Journal.*

[38] Pensacola Can Have Airmail To North Now. (1928, April 7). *The Pensacola Journal.*

[39] New Airmail Route Causes Map To Shrink. (1928, May 2). *The Pensacola Journal.*

[40] Airmail Needs Better Highway. (1928, April 29). *The Pensacola Journal.*

[41] Pensacola Now Uses Airmails. (1928, April 20). *The Pensacola Journal.*

[42] Writers Don't Use Airmails. (1928, May 17). *The Pensacola Journal.*

[43] Record Flood Menaces West Florida. (1929, March 15). *The Pensacola Journal.*

[44] City's Letters Ride In Planes As Waters Rise. (1929, March 16). *The Pensacola Journal.*

[45] Airmail Gains Through Flood. (1929, March 21). *The Pensacola Journal.*

[46] Airmail Gains Through Flood. (1929, March 21). *The Pensacola Journal.*

[47] Airmail Gains Through Flood. (1929, March 21). *The Pensacola Journal.*

[48] Lions Plan On Airmail Route For Pensacola. (1929, September 19). *The Pensacola Journal.*

CHAPTER 20
FAVORITE SONS

Figure 1. From the left, and barely visible in this April 1, 1923 photo, are Gordon Smedly, advertising manager for *The Pensacola Journal*, Bryan Mack, editor of *The Pensacola Journal*, Lieutenant Harry Thompson, and Joseph Browne, business manager of *The Pensacola Journal*. (Courtesy *Pensacola News Journal*)

FAVORITE SONS

Affectionately known as the "Mother-in-law of the Navy," Pensacola laid claim to thousands of Naval aviators by birth, marriage, or their tours of duty in Pensacola. Yet of all these favorite sons, three aviators were honored with particular admiration and respect. And the most prominent was not even a Naval aviator, but an Army aviator born in Milton.

HARRY W. THOMPSON

One of the greatest pioneers of commercial aviation in northwest Florida, Harry W. Thompson was born in Milton, where his father was clerk of the circuit court of Santa Rosa County.[1] He graduated from Milton public schools before enrolling at Boston Tech and subsequently joining the U.S. Army Air Service. He first appeared in *The Pensacola Journal* in 1917 when he was sent by the U.S. Army to Pensacola to study water aeronautics.[2]

In May 1918 he flew to Pensacola in support of the Third Liberty Loan campaign. Piloting a JN-4D from Memphis, Tennessee, he put on several aerial demonstrations over his hometown of Milton and Pensacola. Later that

Figure 2. Harry W. Thompson of Milton served in the U.S. Army Air Service and was one of Pensacola's strongest civil aviation supporters. (Courtesy *Pensacola News Journal*)

month, Thompson was injured in a serious airplane accident at Park Field that resulted in a broken leg, collarbone and jaw.[3] In November 1918, the paper reported that he had recovered, was serving as a cross-country instructor, and expected to be discharged from the U.S. Army soon.[4]

By April 1919, Thompson was out of the service, serving as chairman of the speakers bureau of the Liberty Loan Committee and arranging exhibitions in support of the drive in Pensacola.[5] In 1920, he was heavily involved in the establishment of Kiwanis Field, Pensacola's first commercial landing field.

In June 1922, he was featured prominently at a Kiwanis Club luncheon for Admiral Moffett. He delivered a brief speech in which he:

...called attention to the great importance of using every effort to retain the Naval Air Station here, and read some figures to show the large amount of money which is

---

MOTHER-IN-LAW OF THE NAVY

The opening of the Naval Aeronautic Station in 1914 introduced the dashing young naval aviator into Pensacola's society. One young lady noted that "Naval aviators ...had it all aver the town's men. They had to pass strict physical examinations and were physically fit. They also had been around quite a lot - danced well, played golf and bridge and as Annapolis men had to be topnotch."

Another writer commented that "The flier's combinatioin of charm, adventure, handsome uniform and financial security proved irresistable" and that "Since the Navy boys were excellent marital prospects, parents were inclined to be permissive in their daughter's relationship with them."

The local newspaper noted that "Only one graduating class of fliers between 1914 and 1923 left Pensacola without taking along one of its young women as a wife." By 1931 it was estimated that more than one hundred Pensacola women were married to Navy and Marine Corps officers, and an October 25, 1931 newspaper article listed them all by name!

McGovern, J.R. (1976). *The Emergence of a City in the Modern South: Pensacola 1900-1945.* DeLeon Springs, FL: E.O. Painter Printing Company

spent in Pensacola by the men of the Navy. Mr. Thompson said that from figures he had secured from the commandant's aide at the Pensacola station, he found that for the last three months the payroll for civilians at the air station has amounted to $60,000 a month, with the enlisted personnel receiving $59,000 per month and the officer personnel, $39,000, making a total of $154,000, or $1,084,000 annually.[6]

In April 1923, he appeared in a front-page photograph in *The Pensacola Journal* standing in front of a DH-4. He had flown 200 miles from Montgomery, Alabama to Pensacola in one hour and twenty minutes to transport some desperately needed newspaper engravings for *The Pensacola Journal*.[7] In this article, he also was identified as the vice president of the Gulf Coast Air Line.

In October 1927, his name appeared again in *The Pensacola Journal* as the chairman of the Crestview Legion post that sponsored the development of the airport at Crestview.[8]

He eventually relocated to Tallahassee where he served as assistant state service officer. He passed away at an early age on December 17, 1931.

J. ALBERT WHITTED

Born on February 11, 1893 in St. Petersburg, Florida and raised there, J. Albert Whitted enlisted in the U.S. Navy and reported to Pensacola for flight training. He married a Pensacola girl, Frances Brent, and with her had two small daughters, Catherine and Frances.[9]

Upon his release from active duty in 1919, he purchased a surplus Curtiss two-person flying boat from the U.S. Navy. He christened it the *Blue Bird* and established the first commercial seaplane business in Pensacola at the rear of the Pensacola Yacht Club on Palafox Street.[10] While he would have several aviation businesses elsewhere, he often returned to Pensacola for various flying ventures.

Figure 3. Pensacola's first commercial aviation pioneer was former Naval aviator James Albert Whitted. (Courtesy of Homan & Reilly Designs)

On July 11, 1920 *The Pensacola Journal* reported that Whitted had been added to their staff to carry papers to Camp Walton[a] and other resort areas. *The Pensacola Journal* stated that it was the "...first time that a Florida newspaper has been delivered by airplane. It marks an epoch."[11]

---

[a] A popular resort area 40 miles west of Pensacola now known as Fort Walton Beach.

In February 1922, Whitted launched

...his new seaplane 'Falcon' at the Barcelona street wharf. The plane, which is known as a two-seat type, was built by Lieutenant Whitted after his own design, the work being practically completed by himself. It has a wing spread of 50 feet with a four-blade propeller. The hull is of mahogany, wings painted buff. It contains two cockpits capable of carrying three persons each, the plane thus carrying five passengers besides the pilot.[12]

The same article also noted that

Lieutenant Whitted in all his experience as a flyer, in which he has taken several thousand passengers up in the air, has never met with a mishap and is considered one of the most able pilots. He does no stunts and takes no chances while flying.[13]

Figure 4. James Albert Whitted's headstone at St. Michael's Cemetery in Pensacola. (Courtesy of author)

He departed Pensacola on February 26, 1922 for St. Petersburg in *Falcon* on a trip that he estimated would take from five to six hours to cover the distance of over 400 miles.[14] He told *The Pensacola Journal* that "...he expected to spend the next few months at his home in St Petersburg, but was later going to take an extensive trip up the Atlantic Coast.[15] Whitted was back in Pensacola in July 1923, flying for Gulf Coast Air Line.[16]

On August 19, 1923 tragedy struck. Whitted was in *Falcon* with four passengers on a trip from Camp Walton when:

Flying at an altitude of about 200 feet, the propeller of the plane tore apart, cutting the control wires and partly cutting off the rear portion of the fusilage (sic). The machine dropped into the sound, 20 feet from the island, in 12 feet of water. All occupants were killed instantly. A part of the four-blade propeller was whirled across the sound, more than 300 yards, onto the mainland. Besides residents along the shore opposite where the plane fell, a number of persons along the waterfront in front of The Harbeson Hotel, where Whitted and his four passengers had taken off ten minutes before, heard the mtor (sic) when it stopped suddenly and heard the resounding crash of the heavy machine when it struck.[17]

Pensacola had lost one of her dearest sons. He is buried in St. Michael's Cemetery in Pensacola and Albert Whitted Airport at St. Petersburg is named in his honor. His death also sparked a surprising editorial in *The Pensacola Journal* over the "...necessity of government inspection of commercial airplanes, and the licensing of pilots."[18]

NOEL DAVIS

Lieutenant Commander Noel Davis was born in Salt Lake City, Utah, earned his Naval aviator wings in Pensacola in August 1921, and married a Pensacola girl, Mary Elizabeth Merrit.[19]

In April 1927 he was one of a group of aviators, including Charles Lindbergh, who were competing for hotel owner Raymond Orteig's prize of $25,000 for the first aviator to fly non-stop between New York

Figure 5. Noel Davis' headstone at St. John's Cemetery in Pensacola. (Courtesy of author)

and Paris. He endeared himself to Pensacolians when he announced that he might start his trans-Atlantic flight to Paris from Pensacola. He stated that he "...considered this city the best place to start from because he can get the benefit of warm, moderate winds until his plane is far out in the Atlantic."[20]

While conducting his final test flight on April 26, 1927 at the maximum fuel weight required for the transoceanic trip, his aeroplane the *American Legion* crashed shortly after takeoff. Davis and his co-pilot Lieutenant Stanton Wooster, U.S. Navy, were both killed. [21]

An article in *The Pensacola Journal* described his funeral in Pensacola: "Noel Davis came home yesterday. Back to the city in which he received his training as an aviator, and in which he was married, the body of the distinguished pilot, who was killed in a crash of his plane in Virginia Tuesday, was brought for funeral services and interment."[22]

OTHER AVIATORS

Figure 6. Headstone of Enoch McIntosh, Chief Aviation Pilot, US Navy, at St. John's Cemetery. He was killed in a mid-air collision above Corry Field. He left behind a wife and a small daughter. (Courtesy of author)

Other aviators not quite so famous joined their brothers-in-arms in Pensacola cemeteries. Killed in Pensacola or at other Naval air stations around the nation, their broken bodies would be returned here to their final resting place. Typically, a brief article in *The Pensacola Journal* described their fatal accident followed later by a short note that described the funeral. Invariably these aviators left behind a wife and young children and one cannot help but wonder how they coped with the sudden tragic loss of their husband and father.

After Noel Davis' death, an editorial in *The Pensacola Journal* honored these aviators:

Although Noel Davis goes to join an array of aviators who have died in the attempt to prove the practicality of flying, his death will not deter those left behind from trying to do what he had set his mind upon accomplishing. Instead fatalities inspire them to prove that the feats, which were planned, can be accomplished under favorable conditions. Zachary Landsdowne, John Rodgers, Noel Davis, and the others, beckon the unafraid birdman to keep on because they made the supreme sacrifice. Some day the craftsmanship of the planes will equal the spirit of those who will fly them across the ocean, and men like Noel Davis will not have lived in vain. [23]

NOTES

[1] U.S. Is Calling For Pilots In Aerial Mail. (1918, November 27). *The Pensacola Journal*.

[2] Milton Boy In The Aircraft Service. (1917, October 20). *The Pensacola Journal*.

[3] U.S. Is Calling For Pilots In Aerial Mail. (1918, November 27). *The Pensacola Journal*.

[4] U.S. Is Calling For Pilots In Aerial Mail. (1918, November 27). *The Pensacola Journal*.

[5] Flying Will Feature Loan Drive Today. (1919, April 28). *The Pensacola Journal*.

[6] More Planes Likely For Pensacola Naval Air Station. (1922, June 16). *The Pensacola Journal*.

[7] Thompson Makes Speedy Hop To Bring Pictures. (1923, April 1). *The Pensacola Journal*.

[8] New Air Field At Crestview Ready For Use. (1927, October 19). *The Pensacola Journal*.

[9] Five Killed In Crash At Camp Walton. (1923, August 20). *The Pensacola Journal*.

[10] Pensacola Has First Chance To Enjoy Hop. (1919, October 22). *The Pensacola Journal*.

[11] Journal Delivery Is Made By Plane. (1920, July 11). *The Pensacola Journal*.

[12] Successful Flight Of Whitted's Plane. (1922, February 20). *The Pensacola Journal*.

[13] Successful Flight Of Whitted's Plane. (1922, February 20). *The Pensacola Journal*.

[14] 'Falcon' Will Fly To St. Petersburg. (1922, February 26). *The Pensacola Journal*.

[15] Falcon' Will Fly To St. Petersburg. (1922, February 26). *The Pensacola Journal*.

[16] Journal Runs Extra On Shelby Scrap: Quick Delivery By Seaplane And Auto. (1923, July 5). *The Pensacola Journal*.

[17] Five Killed In Crash At Camp Walton. (1923, August 20). *The Pensacola Journal*.

[18] Government Inspection Of Airplanes. (1923, August 21). *The Pensacola Journal*.

[19] Pensacola's Military/Naval Aviation at St. John's Historic Cemetery. Retrieved from http://www.stjohnhistoriccemetery.com/pensacola_heritages/naval_heritage.

[20] Pensacola-Paris Flight Planned. (1927, April 25). *The Pensacola Journal*.

[21] Forrest, L.C. (1999, March-April). Those Willing To Dare. *Naval Aviation News*, pages 22-24.

[22] Body Of Noel Davis Brought Here For Funeral And Burial. (1927, April 30). *The Pensacola Journal*.

[23] Price of Progress. (1927, May 2). *The Pensacola Journal*.

CHAPTER 21
CHARLES A. LINDBERGH VISITS PENSACOLA

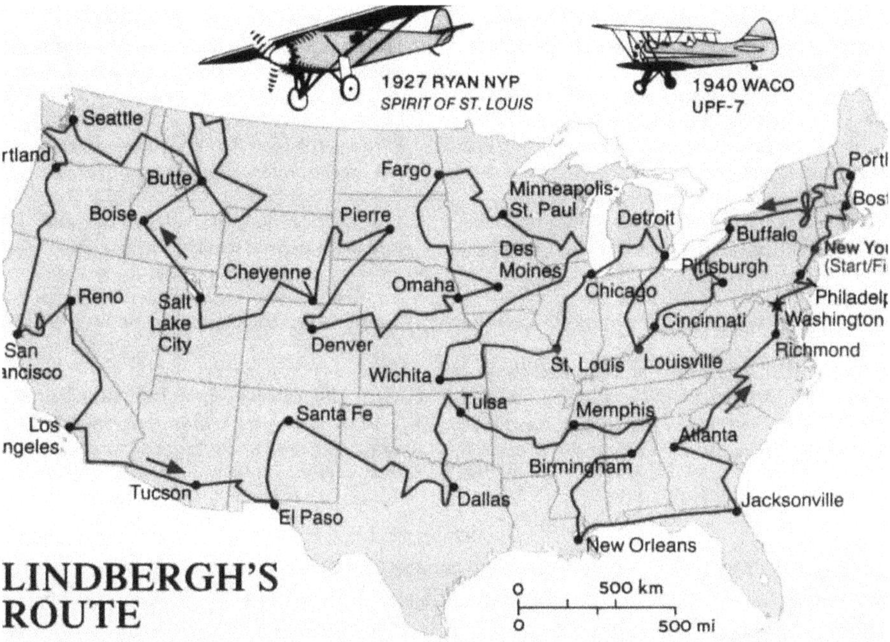

1927 RYAN NYP
SPIRIT OF ST. LOUIS

1940 WACO
UPF-7

LINDBERGH'S ROUTE

Figure 1. Only two months after completing his historic flight across the Atlantic Ocean, Charles Lindbergh departed on a tour of the United States in the *Spirit of St. Louis* to promote commercial aviation. Beginning in New York, he made 82 stops in 48 states in 95 days, including a non-scheduled visit to Pensacola. (Used with permission of the *National Geographic Society*)

LINDBERGH CRASHES IN PENSACOLA

Charles Lindbergh first flight in a plane was on April 9, 1922 as a 20 year-old flight student with the Nebraska Aircraft Corporation. After about eight hours of instruction, he started barnstorming[a] around the United States until applying to become a "Flying Cadet" in the U.S. Army Air Service.[1] He was accepted and ordered to report to Brooks Field at San Antonio, Texas in March 1924. He departed Lambert Field, Illinois on January 23, 1924 with a friend named Leon Klink, who owned a Canuck JN-4C. They intended to barnstorm their way down to Brooks Field.

Along the way they stopped at Station Field aboard Naval Air Station Pensacola "...where the Commanding Officer showed us every courtesy during our visit."[2] In his biography he described what happened upon their departure:

---

[a] In his book *We*, Charles Lindbergh described barnstorming as an "...aviator's term for flying about from one town to another and taking any one who is sufficiently 'air minded' for a short flight over the country. In 1922 the fare usually charged was five dollars ($48.49 in 2002 dollars) for a ride of from five to ten minutes."

Klink and I decided to cut short our stay at Pensacola and to work our way as far west as time would allow before it was necessary for me to leave for Brooks Field.

We had promised to take one of the ladies of the post for a short hop before leaving, and on the morning of our departure I took off for a test flight before taking the lady over Pensacola. Just after the ship had left the field and was about two hundred feet high over the bay, the motor 'reved' down to about five hundred. I banked around in an attempt to get back to the field but lacked by about fifteen feet enough altitude to reach it, and was forced to land in the sand hills less than a hundred feet from the edge of the flying field. The first hill wiped off my landing gear and one wheel went up through the front spar on the lower left wing, breaking it off about two feet from the fuselage.

A quick survey of the plane showed that we would require a new landing gear and propeller in addition to the material required to splice the spar.

The Navy hauled the plane into one of its large dirigible hangars and allowed us to make use of its equipment in repairing the damage. We purchased a spare landing gear and a propeller, then built a box frame around the broken spar and after gluing all the joints, screwed it in position and wound the splice with strong cord, which was then shrunk tight by several coats of dope. In this way the splice was made stronger than the original spar had ever been.

When we were not working on the ship we made several trips to the old Spanish forts which protected the city during the days when Florida still belonged to Spain. These are in an excellent state of preservation and contain a number of passageways, one of which is supposed to lead underground between the two fortifications, but although we searched carefully for the opening to this tunnel we never found it.

In all we spent about a week repairing the plane and when it was ready to fly once more I tested it with an Irving parachute borrowed from one of the officers of the station. That was the first service type of chute I had ever worn and I experienced the unique feeling of not caring particularly whether the ship held together during the tests or not. I put that Canuck through maneuvers which I would never have dreamed of doing with it before, yet with the confidence of absolute safety.[3]

Still an unknown, all of this occurred three years before Lindbergh made the first non-stop solo crossing of the Atlantic Ocean in the *Spirit of St. Louis* in 1927.

ARTHUR E. FORSTER RECALLS THE 1924 VISIT

Arthur E. Forster, who worked at Naval Air Station Pensacola at the time of Lindbergh's first visit, recalled the incident:

One day a fellow flew in that had been barn-storming, as they called it. He landed on the station…(and)…went to the Commandant's office, told the Commandant of his plight. The commandant called my boss Lt. Tucker. Lt. Tucker called me and I went up in the office and met this fellow Lindbergh. He was a tall fellow, I think if you say six foot six you are still too short. He was a tall boy, nice fellow to talk to. And he wanted some lower wings and some struts and god knows what else he didn't want.

But it was the Commandant's order to my boss to let him have what he wanted. He said he had to encourage these barn-stormers, because it helps to maintain aviation and that's why we're in the aviation business.

There was a young sailor on the station. I think he was over in one of the squadrons over on the landing field, at that time, I think, his name was (sic) or something like that and he and Lindbergh became good friends. Might have been that they knew each other before Lindbergh came here, that I don't know. But I know Lindbergh was here a week or so and this fellow had Lindbergh stay at his home during the time he spent in Pensacola. He was here a week or more, more or less I don't know. Now this sailor and Lindbergh brought the pieces (sic) over to our shop that they wanted to either repair or exchange. Now we had so much of that material around that there was no need repairing any of it. We wasn't repairing any of it, we were just drawing spare parts. So we managed to look thru the stack of wings that we had and he picked out a pair of wings they were shop worn, but brand new. Been in storage all during the war and never took their place on an airplane. So those were the wings that we gave Lindbergh.

He took those wings, probably some of the struts, landing gear parts, tail skags and quite a few other pieces. He left Pensacola with a good airplane, just as good as any of them we were flying around on the station at that time.[4]

AVIATION RIGGER AND MRS. LADEBAUCHE

While Lindbergh was repairing his aeroplane, Aviation Rigger and Mrs. Edward E. Ladebauche invited him to stay with them in their home in Woolsey. After the repairs were completed, Lindbergh and Kline continued their barnstorming journey to Brooks Field, but not before he took Mrs. Ladebauche for a flight before their final departure from Pensacola.[5] The Ladebauches kept in touch with their guest over the years and their hospitality was returned years later when Lindbergh was a little more famous.

FLYING TOUR OF THE UNITED STATES

At 7:52 a.m. on May 20, 1927 Lindbergh took off in the *Spirit of St. Louis* from Curtiss Field, Long Island for Paris, France. Thirty three hours and 39 minutes later he landed at Le Bourget Airfield, completing the first solo non-stop flight across the Atlantic Ocean.[b] Overnight he became an international phenomenon. On June 11, 1927 Lindbergh returned to the United States aboard the USS MEMPHIS to a hero's welcome.

In an effort to stimulate interest in commercial aviation, the Daniel Guggenheim Fund for the Promotion of Aeronautics[c] sponsored a tour of

---

[b] Between August 18 and 19, 1932, James A. Mollison flew from Portmarnock Strand, Ireland to Pennfield Ridge, Canada in his DeHavilland Puss Moth *Heart's Content* to make the first solo non-stop westward flight across the Atlantic Ocean.

[c] Daniel Guggenheim established this fund in 1926 with a $2.5 million grant. His son, Henry, was Naval aviator Number 1129. Some aviation historians consider this fund to have restored the United States to a position of world aeronautical leadership.

Figure 2. Charles Lindbergh visits NAS Pensacola during his United States tour following his successful crossing of the Atlantic Ocean. (Courtesy U.S. Navy)

the United States by Lindbergh in the *Spirit of St. Louis*.[6] On July 20, 1927 Lindbergh took off from Mitchel Field in New York to visit 82 cities in 95 days in each of the then 48 states. Pensacola was not a stop on this 22,350-mile trip but it did include overnight stops in nearby Jackson, Mississippi and New Orleans before proceeding on to Jacksonville.

When he flew into Jackson, Mississippi the Ladebauche family and several representatives of Pensacola, including some Naval aviators, were there to watch him land. As described in the paper:

When Lindbergh reached the Jackson landing field, among the crowd gathered to meet him he saw a familiar uniform and a familiar face. Just outside the enclosure stood Aviation Machinist's Mate Ladebauche in his dress blues. Leaving for a moment the dignitaries who were escorting him, Colonel Lindbergh hastened to his old friend with a smile and handclasp and a few words of which the most important were "Come to see me." The invitation was repeated at the banquet that night when a note requested Ladebauche to call at the Colonel's hotel immediately after the banquet.[7]

After a warm reunion that evening and encouraged by the Ladebauches to visit Pensacola, Lindbergh generously agreed to visit Pensacola for an informal call once he was in New Orleans, where a rest day was scheduled. The graciousness of this offer cannot be overstated since his official itinerary established by the Daniel Guggenheim Fund did not permit even the slightest variation in schedule.[8] The *Spirit of St. Louis* would have to remain in New Orleans due to contract requirements but the U.S. Navy offered one of their aeroplanes in its place. The Navy pilot who would fly with Lindbergh from New

Figure 3. A copy of the letter Charles Lindbergh dropped was published in the *Air Station News*. (Courtesy U.S. Navy)

Orleans to Pensacola was none other than First Class Seaman Ladebauche.[9] In preparation for his visit, *The Pensacola Journal* noted:

...owing to the impossibility of parking the large number of cars that would otherwise go to the station, no automobiles will be admitted within the gate. However, pedestrians will be admitted. If the colonel flies to Corry field, the public is urged not to crowd upon the field and perhaps make it impossible for him to land or take off.[10]

On October 9, 1927 Lindbergh flew a U.S. Navy Curtiss Hawk provided for his use from New Orleans to Pensacola Naval Air Station.[11] He spent the evening here, where he and Mr. and Mrs. Ladebauche were guests of honor at a dinner at the station.

Upon returning to New Orleans after his visit to Pensacola,

Figure 4. On October 6, 1930, almost three years exactly after Lindbergh's visit, the first aviators to fly east-to-west across the Atlantic Ocean landed at Pensacola. They were the Frenchmen Dieudonne Coste (left) and Maurice Bellonte who made their crossing from Paris to New York on September 1-2, 1930. Several thousand Pensacolians watched as they landed their aeroplane *Point d'Interrogation* (*Question Mark*) at Old Corry Field for a brief stop on their goodwill tour of the United States made immediately after their historic flight. (Courtesy *Pensacola News Journal*)

Lindbergh boarded the *Spirit of St. Louis* and flew eastward to make his scheduled arrival at Jacksonville. Enroute he circled over Pensacola and dropped a canvas bag with an orange streamer attached to attract attention"...which was picked up at Palafox and Zarragossa street and taken to city hall."[12] Although a preformatted message like others dropped at 192 cities not scheduled for a visit, it still impressed the citizens of Pensacola.

T.H. "DOC" KINCAID VISITS

Seventy six years removed from Lindbergh's crossing of the Atlantic, it is difficult for us to appreciate just how much of an international hero he was. Anything associated with Lindbergh generated excitement. Even the 1928 visit of T.H. "Doc" Kincaid "...the man who tuned up the motor on the aeroplane of Charles A. Lindbergh before his epoch-making flight to Paris..." caused a sensation in Pensacola.[13] Serving as a representative of the Wright Aeronautical Corporation, he was at the Naval Air Station to inspect motors. During an interview he noted:

Pensacola has the jump on most sections of the country...You have one of the finest equipped, if not the finest, air stations in the world. Your city should begin plans for providing for future traffic in the air.[14]

"GREAT 1927 NEW YORK TO PARIS AIR DERBY"

Few people realize just how many aviators were vying to be the first to fly non-stop between New York and Paris. In the eight months prior to Charles Lindbergh's successful crossing, six aviators were killed preparing for the same attempt.

On September 21, 1926, Jacob Islamoff and Charles Clavier were killed and Rene Fonck and Lawrence Curtin injured when their Sikorsky *S-35* crashed on takeoff at Roosevelt Field, New York.

On April 16, 1927 Richard Byrd, George Noville, Floyd Bennett and Anthony Fokker were injured when their Atlantic Aircraft Corporation Fokker *America* crashed on a test flight.

On April 24, 1927 Charles Levine and Clarence Chamberlin escaped injury when their Bellanca *Columbia* was damaged on a demonstration flight.

On April 26, 1927 Noel Davis and Stanton Wooster were killed when their Keystone Pathfinder *American Legion* crashed during a test flight at Langley Field in Virginia.

On May 8, 1927, Charles Nungesser and Francois Coli were killed when their Levasseur PL-8 *White Bird* vanished during a Paris to New York flight attempt.

Of interest, Byrd and Chamberlin repaired their aeroplanes and were ready to cross the Atlantic before Lindbergh. They were both waiting for better weather when Lindbergh decided to launch.

Less than a month after Lindbergh's successful flight on May 20-21 1927, they both crossed the Atlantic. On June 4, 1927 Charles Levine and Clarence Chamberlin successfully flew to Germany, although they crashed after running out of fuel in heavy weather. On June 29, 1927 Richard Byrd, Bert Acosta, Bernt Balchen, and George Noville successfully flew to France but ditched offshore, also due to bad weather.

Jablonski, E. (1972). *Atlantic Fever*. New York: The MacMillan Company.

NOTES

[1] Lindbergh, C.A. (1927). *We*. New York: Knickerbocker Press.

[2] Lindbergh, C.A. (1927). *We*. New York: Knickerbocker Press.

[3] Lindbergh, C.A. (1927). *We*. New York: Knickerbocker Press.

[4] Forster, A.E. (N.D.). *Story of Lindbergh*. Pensacola Historical Society.

[5] Lindbergh Once Visited Local Naval Station. (1927, May 22). *The Pensacola Journal*.

[6] Keyhoe, D. E. (1928, January). Seeing America With Lindbergh. *National Geographic*, LIII, 1-46.

[7] Col. Lindbergh's Visit. (1927, October 10). *Air Station News, Pensacola, Florida*.

[8] Col. Lindbergh's Visit. (1927, October 10). *Air Station News, Pensacola, Florida*.

[9] In New Orleans. (1927, October 9). *The Pensacola Journal*.

[10] Lindbergh To Arrive Here About 9:30 a.m. (1927, October 9). *The Pensacola Journal*.

[11] Lindbergh To Arrive Here About 9:30 a.m. (1927, October 9). *The Pensacola Journal*.

[12] Lindbergh Reappears Over Pensacola, Drops Message And Speeds On Toward Jax. (1927, October 11). *The Pensacola Journal*.

[13] Famed Mechanic At Air Station. (1928, January 25). *The Pensacola Journal*.

[14] Famed Mechanic At Air Station. (1928, January 25). *The Pensacola Journal*.

CHAPTER 22
MYSTERIES, ADVENTURES, TRAGEDIES, AND DISASTERS

Figure 1. This aeroplane could easily have read "Pensacola to Brazil" when Pensacola was approached by the Wright Aeronautical Corporation to participate in this potentially record-breaking flight. (Courtesy of author)

THE DISAPPEARANCE OF SEAMAN 2ND CLASS E. J. CROWE, USNRF

On December 26, 1917 Seaman 2nd Class[a] Edmund. J. Crowe, USNRF launched in his N-9 seaplane from Naval Air Station Pensacola on a one-hour routine solo training flight around Pensacola Bay.[1] He never returned.[2]

Presumed to have run out of gas after possibly becoming disoriented in a fog bank that developed during his flight, rescuers commenced a search of Pensacola Bay, the Gulf of Mexico and local bayous and swamps. Quite pessimistically, *The Pensacola Journal* surmised that the "...fragile craft was doubtless buffeted about in the heavy seas of the gulf until it was battered to pieces, saving the lone occupant a horrible death from exposure and starvation."[3]

Several false leads were followed, including one report of an airplane "... barely grazing the treetops, apparently afire and beyond control..." reported near Beck's Lake, 21 miles northeast of Pensacola.[4] Seaman Crowe's uncle and sister traveled to Pensacola to assist in the search, but they eventually departed brokenhearted, with no sign of their loved one. Finally, after several days of fruitless effort, the search was abandoned and the young man presumed dead. *The Pensacola Journal* reported that a radius of 150 miles from Pensacola was thoroughly searched, including sending boats and harbor tugs inland "...until nearly every square mile of West Florida territory had been covered."[5]

---

[a] Rank assigned to student officers who enlisted under the United States Navy Reserve Force.

Then, nearly nine months later, on September 13, 1918 there was a gruesome find:

A fisherman from Pensacola, who was rowing down the Perdido river noticed what he believed to be a portion of a hydroplane sticking out of the mud in the swamp along the river about ten miles above Perdido Bay. Investigation proved that it was a hydroplane and that there were portions of a human skeleton amid the ruins of the plane.[6]

Seaman Crowe had been found. The story went on to note:

The fact that only a portion of the wrecked plane has been found and the skeleton was badly broken leads the Naval Air Station authorities to believe that the young man was wrecked further up the river and that the resting place of the plane was discovered was one brought about by the tide and storms which constantly change the swampy water courses of the Perdido river.[7]

Then in March 1924, over six years after Seaman Crowe was reported missing, *The Pensacola Journal* reported an even more horrifying discovery:

SKULL OF AIRMEN KILLED LONG AGO BELIEVED FOUND

Pensacolian Finds Skull in Perdido Swamp Thought To Be That of Crowe

What is believed to be the skull of Lieutenant Crowe, student aviator of the Pensacola Naval Air Station who was killed in 1917 when his N9 plane crashed into the Perdido swamp after running out of gas, has been found by Grady Fillingim in the swamp about four miles north of Muscogee.

Lieutenant Crowe was killed December 26, 1917, when he left the station to cruise over what was known as the northern course around Pensacola bay. As he turned the nose of the plane southwestward to return to the station he became enveloped in a fog and lost all sense of direction.

Flying around in the fog in an effort to locate the station, the aviator's gas supply was exhausted, it was believed, and in the resulting fall he was killed.

Several years later the demolished plane and fragments of decayed clothing identified as that of Crowe were found by hunters. Continued search by planes and speed boats over the gulf failed to locate the missing airman.

The skull was shown by Mr. Fillingim to several visiting cousins and was brought to The Pensacola Journal yesterday, where it may be seen by anyone interested in it.[8]

SPECTACULAR NIGHT DIRIGIBLE FIRES

One of the more sensational incidents involving dirigibles occurred the evening of September 30, 1918. It is interesting to compare both *The Pensacola Journal's* article and the official U.S. Navy report. *The Pensacola Journal* article:

DIRIGIBLE BURNS IN MID-AIR AT LOCAL STATION

The explosion of a dirigible at Woolsey and its flaming flight over the bay occasioned much alarm in Pensacola and Warrington last night about 10:20 o'clock.

It was rumored on the street that the machine caught fire in the air and exploded, causing the death of five men.

This was denied at the naval air station, the official information being that it was not in flight when the accident occurred and that no serious injuries resulted.

It is stated that the dirigible was preparing for a flight when it caught fire and that the flames forced the men in charge to release it.

As the flaming machine sailed upwards its spectacular flight was witnessed by people in all parts of the city and many exaggerated rumors resulted.

The men who were handling the machine were just about to make a flight and some burns were sustained by them, forcing them to release the 'blimp' which sailed out over the bay, its spectacular flight creating panic in Pensacola, from which it could be plainly seen.

Residents of Warrington stated that two men were seen to jump from the machine, but officials said that what was seen was probably the falling of the engines.[9]

The official U.S. Navy report:

At 10:10 p.m. September 30, 1918, Goodrich dirigible A-245 was about to start on a flight when the fuselage took fire because of a back fire in the intake manifold of the auxiliary blower engine. The force of men holding the balloon down by the grab ropes, aware of the danger from the great volume of hydrogen gas in the balloon, deserted their posts and the dirigible shot upward, giving the four men in the fuselage barely time to jump; the last one fell about thirty feet and sustained only cuts and bruises. Another man was burned about the face and hands. Carried by a northwest wind, the balloon traveled rapidly across the station and over Pensacola bay where at a height of one thousand feet it burst into flames with a dull report and was totally destroyed. It presented a truly impressive spectacle with the tremendous mass of flame high in the air with burning parts of the fuselage falling, especially terrifying because the onlookers except the crew that let the balloon go, supposed that the fire had started in the air and that there were men in the flames.[10]

In another incident on November 25, 1918 kite balloon A-5029 was "riding about 800 feet high to a mooring line (wire) for the purpose of an endurance test of its fabric, without passengers (when it) was struck by lightning and fell in flames, totally destroyed."[11]

Evening entertainment for Pensacolians at the expense of the U.S. Navy!

FLYING INTO A HANGAR

On June 14, 1918 Lieutenant (Junior Grade) H.S. Alden, flying a N-9 seaplane, made one of the most sensational landings in Naval Air Station Pensacola's history. Approaching the station for landing, he:

...skidded over Warrington beach and flew directly into the portal of a hangar which gave away and the struts and wing members lodged themselves in the structural steel

so that Lieutenant Alden and his plane were suspended at least twenty feet from the ground. The only part of the plane which was not badly broken up was the rear cockpit in which Lieutenant Alden was seated, and the extent of his injury was a slight laceration of the face. It might be added that it was necessary for Lieutenant Alden to descend from his plane by means of a ladder.[12]

Figure 2. For the filming of the 1957 movie *Wings of Eagles,* stunt pilot Paul Mantz successfully flew an N-9 through a hangar at Pensacola, unlike Lieutenant (Junior Grade) Alden in 1918. (Courtesy U.S. Navy)

FLYING INTO SHIPS

On October 1, 1918 Ensign C.A. Roedell and Landman Machinist Mate E.A. Murpha:

...while rising from the water the plane struck the masts of a fishing schooner, sheared off both spars ten feet from the deck, stripped every bit of rigging from the vessel and fell into the bay. Neither flyer was injured. The aviators had failed to see the schooner as it was between them and a brilliant rising sun.[13]

On October 30, 1928 almost 10 years later to the day, Lieutenant R.R. Darron and Ensign J.M. Gregg did the same thing, crashing into the motorboat *Jewell* during takeoff. Even more remarkable, the seaplane "...stuck in the hole it made in the bow of the motorboat, and was brought to the municipal wharf here."[14]

The *Jewell*'s Captain described the accident:

Skipper And His Wife Miss Flying Death By Few Inches

Captain of Jewell Tells How He Saw Plane Coming But Could Do Nothing To Avert Tragedy

Death roared straight into the faces of a man and woman out in Pensacola bay yesterday afternoon. And they were helpless to do anything but stand, watch and wait. They lived to tell of it last night because that death stopped with a terrific crash less than three feet from where they crouched.

They are Capt. and Mrs. H.E. Houseman of the motorboat Jewell which was struck head on by a seaplane piloted by Lieut. Roy Darron of the naval air station.

Watched Plane

'I saw the plane when it landed on the water about 1500 feet ahead of us.' Capt. Houseman declared in describing the accident.

'I was in the cabin with my wife standing beside me. Suddenly I heard the motor roar when the pilot gave it the gun. It was headed straight for us.

'That plane's going to hit us. I shouted to my wife. She gave one little scream and dodged over on the cot in the cabin. She sure was game. There was no time to run, no time to do anything. I threw the wheel hard over and dropped down behind the wall away from the glass windows.

Awful Crash

'The next instant the plane hit with an awful crash. It took me almost a minute to realize we had not been killed. Then I raised up and looked around.

'The pontoon of the plane had struck the hull on the starboard bow and crashed though two-inch planks. The heavy motor pulling the rest of the plane had sheered off and torn its way across the deck and through the front of the deck house. The propeller was still whirling when it hit and had cut a circle into the front wall.

'The flier in the front seat jumped out and hit the deck almost as soon as the plane did. His first words were 'first aid.' He was bleeding badly.

Sinking

'I was certain we were sinking, and whirled around for the island hoping to reach it and beach the boat.

'About that time, the cabin boy, Eddie Ziel, who was in the deck house when the plane tore its way in, came running out with life preservers. The gas tank of the plane had burst when it hit, and he thought the gasoline thrown over him was water pouring into the boat.

'In a few moments I noticed the boat was not settling, and decided it was not going to sink, so turned around and came on to Pensacola.

Saved By Fate

'The motor and top wing of the plane stopped almost under the spot where my wife and I had been standing. If the plane had struck a foot lower it would have sunk us. If it had struck two feet higher, it would have torn through the front wall and killed both of us.

'Neither of the men in the plane was so badly injured they needed help to get out. They both declared they never saw us until they struck.[15]

DEATH ON THE BEACH

On March 1, 1921 Ensign John Alcorn was piloting an N-9 seaplane over the city of Pensacola. *The Pensacola Journal* described what happened next:

Ensign Alcorn landed at the strip of beach just east of Muscogee wharf at about 12:30 o'clock and picked up a woman and gave her a seaplane ride of about 20 minutes duration. He then landed her on the beach and then went up again. This time he stunted for some minutes and upon coming down to the water apparently side slipped in. His forward speed carried him along the beach at a low altitude for some distance and the three negro women and two negro children were knocked down. Two of the women were killed instantly, one died enroute to the hospital and the other died shortly afterwards. The fifth is expected to die.[16]

Alcorn landed in the water, swam ashore and then immediately reported to the Naval Air Station. He denied to authorities that he had taken a woman flying and stated that he was flying over the city when he had an engine failure and was trying to reach the bay for an emergency landing when the accident occurred. Dozens of witnesses, however, corroborated the story that he had picked up an unknown woman for a flight, and after dropping her off, took off again and then hooked a kite on his tail. He then came down the beach very low hitting the women and children.

The next day a civilian coroner's jury was convened.[17] Fourteen witnesses testified, including Genevieve Schuler, who admitted to riding with Alcorn prior to the incident. This jury cleared the Ensign with a verdict of "unavoidable accident." A Naval Board of Inquiry was convened to investigate the accident but no recording could be found of their findings.

THE MISSING BALLOONISTS

On March 22, 1921 two free balloons launched on an evening training mission from NAS Pensacola.[18] The balloonists' objective was to practice night flying and to cover the 200-mile minimum distance required by such flights.[19] The first balloon landed at Brewton, Alabama the next morning. The second balloon, A-5604 with five men aboard, disappeared.

As recounted in *The Pensacola Journal:*

Shortly after 8 o'clock on Thursday morning, March 24, a carrier pigeon arrived at the air station with a message that the balloon was drifting to sea over Jupiter inlet and was out of supplies. That afternoon a second pigeon arrived with a message saying that the balloon was over the gulf, very low, with drag rope out and drifting northward.

As soon as the first message was received a flotilla of seaplanes was ordered out to search for the balloon, and at the same time an eagle boat and a subchaser were directed to make a search of the surface of the gulf in the vicinity where the balloon was supposed to have been when last reported. These were augmented by a dirigible and additional surface craft, and an advance base was established at Panama City for the purpose of continuing the search.

For more than two weeks the search was continued, but no trace of the missing men was found. Only once did there appear to be even a tangible clue, and that was when fishermen reported hearing cries in the swamps in the vicinity of Lake Wimico. Search of this neighborhood, however, failed to disclose anything.[20]

No trace of the men or balloon was found until two weeks later when a fishing smack found the wreck of the balloon "…floating in the gulf about 50 miles south of Panama City."[21] The balloon was "…almost totally submerged, not more than one-seventh of it being afloat," and "the basket was about six fathoms below the surface, but was still attached to the bag. The balloon was kept afloat by the small amount of gas remaining in the bag."[22]

No trace of the men was ever found but what puzzled naval authorities was the actions of the balloon's senior instructor:

…who was an old and accomplished free and dirigible balloon pilot, did not cut loose the basket and climb with his companions into the load ring in order to gain altitude in an effort to stay in the air long enough to drift to shore. The fact that the basket was still attached to the balloon has deepened the mystery.[23]

The search for the missing men was eventually discontinued and *The Pensacola Journal* noted that "…their fate will always remain one of the mysteries of the air and sea."[24] This was the first balloon or dirigible fatalities at Pensacola.

FLYING INTO A RADIO MAST

After a hurricane struck Pensacola in September 1926, U.S. Army pilot Lieutenant Edwin Johnson of the U.S. Aerial Photographic section at Maxwell Field, Montgomery flew to Pensacola to survey the damage.[25] He returned to Montgomery with a civilian, Pensacola born James M. (Ted) Muldron, Jr. "… in order that the young Pensacolian could get in telephonic communication with his mother and father who were at the Pennsylvania Hotel, New York."[26] Upon the return trip while approaching to land at Station Field Lieutenant Johnson was observed "…zooming near the wireless masts when the plane tore into one of the towers, tearing a wing loose and hurling the plane into the front yard of Commandant Upham's home, where it exploded."[27] Both men were killed.

PENSACOLA TO RIO NON-STOP

On July 6, 1927 the headline "Pensacola-Rio Janeiro Flight Proposed" blazed across the morning edition of *The Pensacola Journal*.[28] Amazed Pensacolians read that Charles L. Lawrence, president of the Wright Aeronautical Corporation, was seeking a pilot to fly non-stop from Pensacola to Rio de Janeiro. The Wright Aeronautical Corporation would provide the airplane and cover all expenses of the flight.[29] If successful, this distance of nearly 4,600 miles would set a world record for the longest non-stop flight completed.

The current record holder was Charles Lindbergh who had flown 3,610 miles across the Atlantic Ocean less than seven weeks before in the *Spirit of St. Louis* powered by a Wright Whirlwind J-5C engine.   Seeking to capitalize

Schedule of Proposed Hop From Pensacola to Brazil

FROM	Distance	Total Distance	Departure	Elapsed Time
PENSACOLA			1:00 p. m.	
FLORIDA KEYS	530	530	6:18 p. m.	5 hr. 18 min.
Jardinese del Rey, Cuba..	210	740	8:24 p. m.	7 hr. 24 min.
Santiago de Cuba	240	980	10:48 p. m.	9 hr. 48 min.
Cape Dame Marie, Haiti ..	150	1,130	12:18 a. m.	11 hr. 18 min.
Willemstadt, Curacao, W. I.	560	1,690	5:54 a. m.	16 hr. 54 min
Caracas, Venezuela	185	1,875	7:45 a. m.	18 hr. 45 min.
Gerualdo, Brazil, (Branco River)	700	2,575	2:45 p. m.	25 hr. 45 min.
Serpo, Brazil (Amazon River)	390	2,965	6:39 p. m.	29 hr. 39 min.
Tapajos River	210	3,175	8:45 p. m.	31 hr. 45 min.
Xingu River	430	3,605	1:03 a. m.	36 hr. 03 min.
Grande River	250	3,865	3:33 a. m.	38 hr. 33 min.
Tocantins River	130	3,985	4:51 a. m.	39 hr. 51 min.
Paranahiba River	240	4,225	7:15 a. m.	41 hr. 15 min.
San Francisco River	160	4,385	8:51 a. m.	43 hr. 51 min.
Rio de Janeiro	210	4,595	10:57 a. m.	45 hr. 57 min.

Figure 3. Table of distances and times from Pensacola to various locations in South America. (Courtesy *Pensacola News Journal*)

on his use of their engine, the Wright Aeronautical Corporation intended to seize this opportunity to expand into foreign aeroplane markets. Lawrence stated that "Our company was already doing some business in South America but these dramatic achievements will cinch the argument concerning the superiority of American products."[30] More specifically, Mr. Lawrence was seeking "...to attract the interest of South American counties to the Wright Whirlwind airplane motor, the type used by Lindbergh, Chamberlin and Byrd on their recent trans-Atlantic flights."[31]

Pensacola immediately recognized the value of such publicity to the city. An editorial noted that "...the *Spirit of St. Louis* and Lindbergh brought more publicity and credit to the city of St. Louis in one 'hop' than all the years of aviation activity that Pensacola has experienced."[32] However, Pensacola had to move quickly for the "...little city of Brunswick, Ga., about half the size of Pensacola, has offered a prize of $25,000 for the first flight from there to Rio de Janeiro."[33]

Mayor J. H. Bayliss promptly sought the services of Captain William A. Lawrence, a former Naval aviator and airmail pilot, by wiring him a telegram.[34] Captain Lawrence eventually declined, but the mayor received many applications from other aviators wishing to pilot the flight.[35]

However, on July 10, 1927 *The Pensacola Journal* reported that Paul H. Redfern, backed by $25,000 raised by a group of Brunswick, Georgia businessmen, had signed a contract to conduct the flight. Redfern announced that "...he would travel alone. His route to the Brazilian capital would be across the Caribbean Sea and over the Amazon jungle and the Andes Mountains."[36]

Redfern would make his attempt in a Wright Whirlwind equipped Stinson-Detroiter monoplane named the *Port of Brunswick* in honor of his city sponsors. He also publicized the fact that he might attempt to break the current world endurance flight record by remaining airborne for more than 52 hours. [37]

At 12:45 p.m. on August 25, 1927 Redfern took off in his green and yellow monoplane from Glynn Isle beach near Brunswick, before 3,000 spectators. Twenty six hours later, on August 26, 1927, at approximately 3:00 p.m. he circled a Norwegian steamship and dropped a note asking the captain to point

Figure 4. The *Pensacola News Journal* published a photograph of a U.S. Navy seaplane flying over flooded land. (Courtesy of *Pensacola News Journal*)

the ship toward land.[38] When this was done, Redfern flew down the length of the ship to establish his bearing, dipped his wings in appreciation, and vanished over the horizon. He was never seen again, another of aviation's great unsolved mysteries.

MISSISSIPPI FLOOD RELIEF

Between 1858 and 1922 the Mississippi River flooded 11 times but the flood of 1927 surpassed them all. Beginning in 1926 severe rainstorms pummeled the Mississippi Valley area for several months. Levees finally broke in April 1927 spilling millions of gallons of floodwaters with devastating results.[39] When the waters finally subsided in August 1927 more than 700,000 people were left homeless with in excess of $1 billion in property damage.

Seaplanes from the Naval Air Station immediately responded and for several months newspaper articles described their adventures in detail. One article in particular stands out:

Aviation has won a place in the future of the Mississippi Valley as a result of exploits during the recent, unprecedented floods and present reconstruction period.

Observations, made from planes on errands of mercy, proved so valuable to levee engineers that it is planned to use them permanently in flood control work. Naval seaplanes still are engaged in charting the topography of the country while flood control work goes forward and their constant use by district engineers has been proposed by officials.

Zoombing (sic) through the skies, hauling food, medicine, relief workers, newspaper men, or doing scout duty for the fleets of relief boats, the aviators made a close study of the vast inland seas stretched for miles beneath them.

Constant peril confronted them, for there were few places where a seaplane could land safely and virtually none for a land ship to alight.

Devising their own charts of the unknown waters, filled with tops of trees and houses and punctuated at intervals with narrow ridges of high land, the fliers gathered information credited with having saved many lives and preventing numerous levees from breaking.

It was the lone aviator, hovering over the frantic scene when a thousand men fought to save the South Bend levee on the Arkansas river, who saw the rampart break and then sped away to drop messages to nearby towns, warning the population to flee from the danger zone. Three other seaplanes had carried hundreds of sandbags from Little Rock and Vicksburg, enabling the engineers to hold the levee several days.

Aviators, skirting the tops of trees as they scanned the rural countryside for marooned persons, brought rescue boats to hundreds in the nick of time.

'In each flight,' says Commander W.G. Child of the Pensacola Naval Air station, who had 38 seaplanes at work in the valley at one time, 'more information was obtained from these aerial surveys and studies than could be gotten in months of using land and water transportation — and this at a savings of thousands of dollars.'

There were few mishaps and only one death was recorded. At the height of the rescue and relief work approximately 65 planes were working in the inundated territory.

A curious sidelight on flying in the zone was presented by snakes which infest the Mississippi lowlands. Commander Child says water moccasins developed a peculiar liking for planes and seldom failed to clamber aboard from the pontoons resting in water.

They hid under cushions or coils of rope and sometimes wrapped themselves about engines to keep warm. Occasionally, they got away on flights only to be dislodged and fall writhing into the water from the height of several hundred feet.[40]

AIRPLANE HORSE TRICKS

While flying in April 1929 Lieutenant Richard M. Oliver, a student aviator, noticed another aeroplane crash and immediately landed at Old Corry Field to telephone the Naval Air Station for assistance.[41] What happened after he reported the crash was related with delight to the readers of *The Pensacola Journal*:

He had left the motor running. He found that in order to get in position for a take-off, he would have to swing the tail of his plane around. He had no help, so he grasped the nether portion of the machine, and started to change its position.

Typical Horse Trick

Here's where the horse part of it comes in.

As the flier took hold of his plane's tail, the motor immediately speeded up, and before Lieut. Oliver knew it, the machine had dashed off across the field, lifted a couple of feet off the ground and landed in a heap.

Now everybody knows that the same thing might have happened to the driver of a high-spirited horse not so very long ago.

How It Happened.

What happened was that Lieut. Oliver, in his anxiety to report the accident had left his control stick well forward on the left hand corner, near the throttle. As he attempted to lift the tail of his ship around, he moved the 'flipper' which moved the 'stick' which struck the throttle and shoved it open.

Just Like A Horse.

No horse ever had so many things to look out for as an airplane, but suppose the driver, entering his buggy, struck the whip and gave Old Dobbin a nudge.

Would he have started off at a run?

Of course he would.

So did Lieut. Olivier's airplane.[42]

NOTES

[1] Bennett, F.M. (1919, March 15). *Historical Data, collection of.* 73-6, Box 2, Special Collections. John C. Pace Library.

[2] Report Lost Plane Found And Returned. (1917, December 28). *The Pensacola Journal.*

[3] Search For Misning (sic) Airplane Abandoned. (1917, December 29). *The Pensacola Journal.*

[4] Naval Plane Seen Near Beck's Lake. (1917, December 30). *The Pensacola Journal.*

[5] Search for Lost Aviator Ends In Abandoned Effort. (1918, January 6). *The Pensacola Journal.*

[6] Skeleton Of Flyer Found By Fisherman. (1918, September 16). *The Pensacola Journal.*

[7] Skeleton Of Flyer Found By Fisherman. (1918, September 16). *The Pensacola Journal.*

[8] Skull Of Airmen (sic) Killed Long Ago Believed Found. (1924, March 31). *The Pensacola Journal.*

[9] Dirigible Burns In Mid-Air At Local Station. (1918, October 1). *The Pensacola Journal.*

[10] Bennett, F.M. (1919, March 15). *Historical Data, collection of.* 73-6, Box 2, Special Collections. John C. Pace Library.

[11] Bennett, F.M. (1919, March 15). *Historical Data, collection of.* 73-6, Box 2, Special Collections. John C. Pace Library.

[12] Bennett, F.M. (1919, March 15). *Historical Data, collection of.* 73-6, Box 2, Special Collections. John C. Pace Library.

[13] Pensacola Naval Air Station Has Had Few Serious Accidents. (1921, May 22). *The Pensacola Journal.*

[14] Seaplane Hits Sound Vessel During Takeoff. (1928, October 31). *The Pensacola Journal.*

[15] Skipper And His Wife Miss Flying Death By A Few Inches. (1928, October 31). *The Pensacola Journal.*

[16] Four Killed As Seaplane Sweeps Beach. (1921, March 2). *The Pensacola Journal.*

[17] Coroner's Jury Clears Alcorn. (1921, March 3). *The Pensacola Journal.*

[18] Five Naval Balloonists Missing from Local Station May Be Lost In The Gulf. (1921, March 25). *The Pensacola Journal*.

[19] Pensacola Naval Air Station Has Had Few Serious Accidents. (1921, May 22). *The Pensacola Journal*.

[20] Fishermen Find Missing Naval Balloon. (1921, April 10). *The Pensacola Journal*.

[21] Fishermen Find Missing Naval Balloon. (1921, April 10). *The Pensacola Journal*.

[22] Fishermen Find Missing Naval Balloon. (1921, April 10). *The Pensacola Journal*.

[23] Fishermen Find Missing Naval Balloon. (1921, April 10). *The Pensacola Journal*.

[24] Fishermen Find Missing Naval Balloon. (1921, April 10). *The Pensacola Journal*.

[25] Prominent Pensacolian Is Killed As Plane Strikes Naval Radio Tower. (1926, September 25). *The Pensacola Journal*.

[26] Prominent Pensacolian Is Killed As Plane Strikes Naval Radio Tower. (1926, September 25). *The Pensacola Journal*.

[27] Prominent Pensacolian Is Killed As Plane Strikes Naval Radio Tower. (1926, September 25). *The Pensacola Journal*.

[28] Pensacola-Rio Janeiro Flight Proposed. (1927, July 6). *The Pensacola Journal*.

[29] Nothing From Lawrence On Flight Offer. (1927, July 10). *The Pensacola Journal*.

[30] Pensacola-Rio Janeiro Flight Proposed. (1927, July 6). *The Pensacola Journal*.

[31] Nothing From Lawrence On Flight Offer. (1927, July 10). *The Pensacola Journal*.

[32] Let's Go After Rio Flight. (1927, July 31). *The Pensacola Journal*.

[33] Let's Go After Rio Flight. (1927, July 31). *The Pensacola Journal*.

[34] Nothing From Lawrence On Flight Offer. (1927, July 10). *The Pensacola Journal*.

[35] Birmingham Aviator Offers To Attempt Non-Stop flight From Pensacola To Janeiro. (1927, July 14). *The Pensacola Journal*.

[36] Brunswick, Ga Signs Man To Attempt Non-Stop Hop To Rio De Janeiro, Brazil. (1927, July 10). *The Pensacola Journal*

[37] Thousands of Georgians See Redfern Plane. (1927, August 8). *The Pensacola Journal*.

[38] Savage, T. & Shelton, R. (n.d.). A Columbia (SC) Aviator and His Stinson Detroiter Remembered. Retrieved from http://www.capnbilly.com/redfern.htm.

[39] Ashley County Ledger. (2002, July 14). *Raging Rivers Wild: A Look at the Flood of 1927*. Retrieved from http://www.ashleycountyledger.com/display/inn_history/Z999.txt.

[40] Work Of Planes Wins Aviation Place In Flood Areas Future. (1927, July 11). *The Pensacola Journal*.

[41] Airplane Plays Horse's Trick On Flier Here. (1929, April 28). *The Pensacola Journal*.

[42] Airplane Plays Horse's Trick On Flier Here. (1929, April 28). *The Pensacola Journal*.

CHAPTER 23
BARNSTORMERS AND FLYING CIRCUSES

Figure 1. A 15 passenger Ford Tri-motor belonging to Tri-Motored Air Tours, Inc. which barnstormed several hundred Pensacolians around their city in November 1928. (Courtesy *Pensacola News Journal*)

FUN IN THE SKIES OVER PENSACOLA

In the immediate post-World War I period, a pilot's career options were few. As noted by one aviation historian: "Without commercial airlines in the United States, American Fliers often had to choose between the hand-to-mouth existence of a barnstormer and the dangerous career of flying the mail."[1] Pensacola was not located on any airmail routes but barnstormers, flying circuses, flying advertisements, and several famous aeroplanes routinely visited the city.

Typically, these aeroplanes operated out of Old Corry Field before crowds of Pensacolians who would either motor out or ride the train to this airfield north of the city. These aeroplanes were also granted access to the facilities at the Naval Air Station for refueling, repairs or maintenance. Not every touring aeroplane was civilian. One of the most famous aeroplanes to visit Pensacola actually belonged to the U.S. Navy.

NC-4

There is no more celebrated aeroplane in Naval aviation history than the NC-4, a Naval Aircraft Factory designed and Curtiss-built flying boat. Under the command of Lieutenant Commander A. C. Read and carrying a crew of six, the NC-4 departed NAS Rockaway, New York on May 8, 1919. On May 27, 1919 the NC-4 landed in the harbor at Lisbon, Portugal to complete the first crossing of the Atlantic Ocean by air in history. After her arrival in England, she was dismantled and shipped back to the United States.

Figure 2. The NC-4 on a seaplane ramp at NAS Pensacola. The NC-4 visited Pensacola at least twice after its unprecedented flight across the Atlantic Ocean. (Courtesy National Museum of Naval Aviation)

In September 1919, the NC-4, with Lieutenant Commander Read and most of his original crew, was sent on a several month long tour of the East Coast, Mississippi River and tributaries, and Gulf Coast in support of U.S. Navy recruiting efforts.[2] She landed in Pensacola on November 2, 1919 for several days of crew rest and an overhaul before continuing on her trip.[3]

The NC-4 returned to Pensacola on December 23, 1919 to spend the holidays here, but not without having stirred up some excitement a few days earlier.[4] Enroute to Mobile, Alabama from Galveston, Texas, the NC-4 was missing for twelve hours.[5] Enveloped in a thick fog, Lieutenant Commander Read decided to land at Grand Isle, Louisana "…because it was shown on the map as a town, but it was destroyed by a tidal wave 25 years ago, drowning 300 persons, and the NC-4 found only a fishing camp and lighthouse in which the officers and crew spent the night."[6]

Soon after completing the tour, the NC-4 was dismantled and its hull acquired by the Smithsonian Institution for display while its engines, wings and tail section languished in several U.S. Navy warehouses. The NC-4 hull was removed from display in 1958 and completely forgotten until Dr. Paul Garber of the National Air and Space Museum took a special interest.[7] Enlisting the support of the U.S. Navy and the Smithsonian

Figure 3. A mechanic at the air station who worked upon the NC-4 recalled that weighing the flying boat out in the open with the wind blowing was "some job." (Courtesy Pensacola Historical Society)

Institution, he coordinated the complete restoration of NC-4 in time for the 50th Anniversary of its flight in 1969, where it was placed on open-air exhibition on the Washington Mall for a single month before returning to storage. Subsequently, the NC-4 was placed on loan to the National Museum of Naval Aviation in Pensacola where it now resides on public display.

MANNA FROM HEAVEN

Companies during this period capitalized upon the popularity of aeroplanes to sponsor various publicity stunts. Whether it was the opportunity to win free aeroplane rides or to catch Baby Ruth candy bars or Peter Pan bread, local businesses would run advertisements in *The Pensacola Journal* hailing these events. The aeroplanes usually were based at Old Corry Field for flights over the city.

FREE
AIRPLANE RIDES

The Saenger management has made arrangements with Capt. Dallas M. Speer, now at Corry Field, in his Baby Ruth airplane, to give 2 free airplane rides to lucky patrons of

HERE'S HOW

Each regular admission ticket bears a number. All patrons attending the show today to see George Bancroft in "The Showdown," should look at their ticket. If it bears No. 078122 or No. 079079, give it to door man with your name and address. He in turn will give you a card good for a ride in the Baby Ruth airplane, piloted by Captain Dallas M. Speer at Corry Field.

Figure 4.   There were no further articles indicating whether either ticket was redeemed. (Courtesy *Pensacola News Journal*)

Peter Pan
Flies Again

Again Peter Pan takes to wings—this time to let you know the goodness of the new loaf which bears his name.

Today, August 12th, at 3:00 in the afternoon, Peter Pan in his special airplane will fly over Pensacola and drop 200 miniature loaves of Peter Pan Bread. Each loaf will be attached to a parachute and to those of you who are fortunate enough to get one of these tiny loaves, your grocer will give you two full sized loaves of delicious Peter Pan Bread.

But whether you get one of these tiny loaves or not, we want you to know the goodness of this fine bread—its crisp Brown crust, its velvety white grain, its different, finer flavor.

Peter Pan is made from a special recipe which has been tested and approved by The Good Housekeeping Institute of Foods, Sanitation and Health. Try a loaf today.

The Peter Pan Bakers
Hughey's Bakery, Inc.

T.L. Gant & Co.

Say:
"PETER PAN" to Your Grocerman

Figure 6. If it's not candy bars, then it's bread being tossed over the sides of aeroplanes! (Courtesy *Pensacola News Journal*)

Figure 5. How can one possibly object to Baby Ruths falling from the sky? (Courtesy *Pensacola News Journal*)

MABEL CODY'S FLYING CIRCUS

The following article appeared in *The Pensacola Journal* in January 1928:

FLYING CIRCUS COMES TO TOWN

Aerial Acrobats Announce Arrival by Performing Stunts over City

Mable (sic) Cody's Flying Circus arrived in Pensacola yesterday. The stunt fliers made known their arrival by performing loops and tail spins over the city. A large number of Pensacolians motored to the old Corry Landing field on Palafox road during the afternoon to witness the stunts.

The fliers, who have four planes, will put on aerial programs this afternoon and Sunday over the city. Mable (sic) Cody is billed to make her change from automobile to plane while both are travelling at high speed.

Yesterday Johnny Gill, Doug Davis, Barney Rowe, Jimmy Krowsky, the 'Flying Dutchman', Frankie Ward and Miss Cody performed numerous stunts. Ward, giving his looping exhibition made 38 loops in succession, 'straightening out' about 300 feet above the ground

Rowe's delayed parachute jump is slated for 4 o'clock today. Wing walking stunts are billed for 3 o'clock. Quite a few persons went aloft for plane rides yesterday and others are expected to do so today.[8]

The Sunday afternoon show apparently went well:

CROWDS WATCH AERIAL STUNTS

Old Corry Field Packed as Mable (sic) Cody's Air Circus Performs

All yesterday afternoon aviators connected with Mable (sic) Cody's flying circus carried Pensacolians on short trips, and towards evening entertained a crowd of several hundred with a series of thrilling aerial stunts.

A feature of the afternoon was the transfer of Miss Cody from an automobile moving at the rate of more than 60 miles an hour to a plane flying over head. The plane which picked Miss Cody up by means of a rope ladder was piloted by Johnnie Gill, former student at the Pensacola Naval air station.

At about 4 o'clock yesterday afternoon, old Corry field on Palafox road was thronged with automobiles which had been flowing onto the field in a steady stream all afternoon. Solid rows of machines were parked near the planes comprising the circus, and special police were required to keep the crowd off the field.

Ride in a
PLANE
1c
per pound
According to your weight

All aboard for **Christmas 1929**

Mabel Cody's Flying Circus

OLD CORRY FIELD

Santa Claus

came in with Doug Davis yesterday in his fast plane. He came to SILVERMAN'S where he distributed Toys to hundreds of good little boys and girls.

GO TO OLD CORRY FIELD AND SEE THE WONDERFUL STUNTS THIS AFTERNOON

Figure 7. Old Corry Field served as host to a variety of barnstormers and flying circuses. (Courtesy *Pensacola News Journal*)

At intervals during the afternoon pilots of the circus planes did stunts overhead. No accidents were reported either on the ground or aloft.[9]

Mabel and her Flying Circus returned to Pensacola in December 1928 with their announcement that "Santa Claus and Doug Davis, airplane racer would arrive on the same plane at Old Corry Field."[10] Santa did land safely and was "...brought by automobiles down town for a parade of the business district... "[11] They were reported to be flying four new Travel Airs, replacing the Wacos flown the previous year at Old Corry Field.

Business was good for rides as more than 400 Pensacolians flew with the Flying Circus and crowds of people witnessed "Dead stick landings, parachute jumping, barrel rolls, tail spins, loops and other stunts..." over the field and city.[12] The Christmas program included Mable Cody attempting "...to transfer from an automobile traveling at 70 miles an hour to a plane piloted by one of the pilots," a stunt she did often.[13]

Figure 8. This 1928 caption read "Doug Davis, crack pilot of the Mable (sic) Cody flying circus, and some of the cups he has won in airplane races all over the east." (Courtesy *Pensacola News Journal*)

---

[a] Now Atlanta's William B. Hartsfield Airport.

One of Cody's pilots, Doug Davis, was a famous aviator in his own right. Born in Georgia, Davis had owned his own flying circus, for which Cody had once worked, and was known throughout the south as the "king of the barnstormers."[14] Davis was also Eastern Airlines' first captain, flew the first commercial airline flight from Atlanta to New York, opened the first hangar on Candler Field,[a] opened Atlanta's first air training school and opened Atlanta's first charter service. In 1929, flying his Travel Air "Mystery Ship" he won the Thompson Cup in the unlimited horsepower free-for-all at the National Air Races. He was killed in 1934 while leading another National Air Race. His statue is in the National Air and Space Museum.[15]

ALL-AMERICA AIR TOUR

Advertised as "…the first time in history, tourists will be able to see almost the entire nation in about a month," Pensacola was listed as one of the cities to be visited in June 1928. Sponsored by American Airway, "25 huge multi-motored planes carrying besides their crews, about 200 passengers, will cover about 10,000 miles."[16] Originating in Baltimore, the planes flew to the west coast and returned to the east coast by a southern route that included stops in New Orleans, Pensacola, Tallahassee, Miami, and Cuba before flying up the Atlantic coast to home.

TRI-MOTORED AIR TOURS, INC.

On November 19, 1928 a barnstorming 15 passenger Ford tri-motor piloted by Captain George Goodsell touched down before a large crowd on Old Corry Field.[17] Owned by the Tri-Motored Air Tours, Inc. of Port Clinton, Ohio, the plane was originally scheduled for only a one-night stay. However, the demand by Pensacolians for barnstorming flights around the city and bay proved so profitable that the plane stayed for several more days.

COMING CHRISTMAS THIS 14 PASSENGER
ALL METAL TRI-MOTOR PLANE

50c Per Ride — WILL BE AT OLD CORRY FIELD —
Tues. 2½ MILES NORTH ON DAVIS **Wed.**
FROM 9 TO 11 O'CLOCK EACH MORNING
60c from 11 to 2; 75c 2 till dark
SEE PENSACOLA FROM THE AIR   SEE PILOT FOR SPECIAL TRIPS

Figure 9. For the enjoyment of Pensacolians, several barnstormers offered Ford Tri-motor rides from Old Corry Field in the early 1930s. (Courtesy *Pensacola News Journal*)

Their initial flight operations were made difficult because "The field was very soft following the hard rain of the morning and considerable delays were experienced as a consequence."[18] However, "new runways have been built at old Corry field to enable the big ship to take off with greater ease. An east and west, and also a north and south runway have been laid out by Captain O.M. Goodsell, pilot, and roled (sic) by city workmen."[19] They also utilized the resources of the U.S. Navy as it was noted that the plane "...stopped at the air station to fill its gas tanks..."[20]

Civilians riders, such as Mrs. Powe, age 68, "Who is spending the sunset hours of her life" in Pensacola commented after her flight that:

I've often stood in the yard at the home and watched the planes flying through the clouds – but I never thought I'd get a chance to fly myself. And now, that I'm up here – and it's so safe and so smooth - I don't blame all those navy boys for wanting to fly. I'll sure have a lot to tell the old ladies at the home, when I get back.[21]

The number of Pensacolians who wanted to fly must have surprised the barnstormers. Apparently, all those U.S. Navy aeroplanes flying overhead every day stimulated their curiosity. On one day alone "Seven flights, carrying more than 100 Pensacolians" were made from Old Corry field.[22]

LOOP-THE-LOOP KING

No less a celebrity than Captain L.A. Thro of Texas Air Lines stopped at Pensacola in September 1929 for repairs while enroute from Jacksonville, Florida to Gulfport, Mississippi.[23] Captain Thro was the current record holder for successive loops, having made 548 of them. When asked what it was like to do so many loops, he replied: "I can't describe it except to say that it gets mighty monotonous."[24]

Figure 10. This photograph found in the West Florida Historic Preservation, Inc. collection may be of exhibition pilot Harold Johnson performing his famous one-wheel landing at Old Corry Field. If not, then some thrill-seeking Pensacolians purchased more excitement than they anticipated on their joyride. (Courtesy West Florida Historic Preservation, Inc.)

NOTES

[1] Josephy, A.M. (Ed. (1962). *The American Heritage History of Flight.* American Heritage Publishing Company, page 232.

[2] Navy's Greatest Plane Is Coming. (1919, November 2). *The Pensacola Journal.*

[3] Famous Plane Lands in Bay for Overhaul. (1919, November 3). *The Pensacola Journal.*

[4] Famous Plane At Navy Yard. (1919, December 23). *The Pensacola Journal.*

[5] NC-4 Expected This Morning. (1919, December 22). *The Pensacola Journal.*

[6] NC-4 Expected This Morning. (1919, December 22). *The Pensacola Journal.*

[7] Smith, R.K. (1973) *First Across!* Annapolis, MD: United States Naval Institute, page 258.

[8] Flying Circus Comes To Town. (1928, January 7). *The Pensacola Journal.*

[9] Crowds Watch Aerial Stunts. (1928, January 9). *The Pensacola Journal.*

[10] Plane Racer Rides Santa. (1928, December 19). *The Pensacola Journal.*

[11] Santa's Planes Arrive In City. (1928, December 22). *The Pensacola Journal.*

[12] Many See Air Circus Stunt. (1928, December 24). *The Pensacola Journal.*

[13] Many See Air Circus Stunt. (1928, December 24). *The Pensacola Journal.*

[14] Stebbins, C. (1989). Doug Davis. *Aviation Quarterly*, 9(2) 132-148.

[15] HR 1445 –Davis, Doug, Sr,: Commend. Retrieved from http://www2.state.ga.us/legis/1999_00/leg/fulltext/hr1445.htm.

[16] Air Voyagers To Visit Here On Long Tour. (1928, March 22). *The Pensacola Journal.*

[17] Huge Airplane Takes Mayor On Trip Over City. (1928, November 20). *The Pensacola Journal.*

[18] Huge Airplane Takes Mayor On Trip Over City. (1928, November 20). *The Pensacola Journal.*

[19] Big Plane To Stay Longer. (1928, November 22). *The Pensacola Journal.*

[20] All Enjoy This Party In The Sky. (1928, November 23). *The Pensacola Journal.*

[21] All Enjoy This Party In The Sky. (1928, November 23). *The Pensacola Journal.*

[22] Huge Air Liner Takes 100 On Trips To Sky. (1928, November 25). *The Pensacola Journal.*

[23] Noted Aviator Stops Here To Repair Plane. (1929, September 7). *The Pensacola Journal.*

[24] Noted Aviator Stops Here To Repair Plane. (1929, September 7). *The Pensacola Journal.*

CHAPTER 24
PIGEON LOFT PENSACOLA

Figure 1. While these pigeon lofts are at NAS North Island, they are probably similar in design to NAS Pensacola's own lofts. (Courtesy of author)

HOMING PIGEONS

We would be remiss in our history of early aviation in Pensacola if we did not mention our fine-feathered friends who also served proudly.

All homing pigeons are believed to be a direct descendant of the wild rock doves of Belgium. Possessing a unique homing ability, special breeding and training enhanced this homing trait to the point where the birds could travel great distances at relatively quick speeds.[1] The use of homing pigeons to carry messages has been traced as far back as ancient Egypt, Greece and Rome. No one has been able to determine exactly how a homing pigeon navigates to its home loft, although several theories have been developed, including magnetic and astronomical navigation theories. Of interest, homing pigeons cannot be used at night or in foggy conditions but are capable of flights of over 500 miles.[a]

PIGEONS AND NAVAL AVIATION

The U.S. Navy's interest in pigeons is traced to the late 1800s when Professor Francis Marion of the U.S. Naval Academy was sent to Belgium to report on their homing pigeons.[2] In 1899, he published a manual on the care and training of homing pigeons and cotes, or pigeon lofts, were established at various shore stations. The U.S. Navy rating Quartermaster (Pigeon), or Pigeoneer, also was established to care and train for the birds.

---

[a] There is no truth to the rumor that scientists attempted to crossbreed homing pigeons with parrots so that the birds could speak their message.

Since early aeroplanes had no radios, aviators quickly realized that homing pigeons provided a means of delivering emergency messages. One Naval aviator remembered his use of pigeons:

Before you started on your flight, you went over to the pigeon loft and got your little box with four pigeons in it. If you had a forced landing, of which we had quite a number, you wrote your message on the piece of paper, stuck it in the capsule that was fastened to the pigeon's leg, and let it go. It flew back to the air station and the people there knew where you were, presumably.[3]

As the development of reliable two-way airborne radio communications was slow, pigeons enjoyed a long and useful life in Naval aviation. As noted by one Navy pilot:

Some of the planes had radio by the early 20s, but we didn't completely trust wireless yet. It was far from reliable. We had to play it safe and take one or two pigeons on each flight. The theory was that if we had to make a crash landing we'd write our location on a slip of paper, place it in a capsule, and fasten it to the leg of our bird. He'd fly back to North Island and alert the rescue unit.[4]

The training requirements of World War I resulted in pigeon lofts being established at fourteen Naval air stations around the United States.[5] More than 600 pigeons were trained at these stations and there were over 2,500 trained pigeons serving overseas during World War I aboard both ships and aeroplanes. In France alone, there were 12 pigeon stations with more than 1,500 pigeons.[6]

After the war, homing pigeons were assigned only to aviation units who used them aboard both heavier-than-air and lighter-than-air craft. Yet even with reliable radios, homing pigeons would continue to be used in Naval viation in lighter-than-air until well into the 1940s.[7] It is difficult to determine when the last pigeons were used in the U.S. Navy but as a reference point, the official date for deletion of the pigeon trainer rating was January 10, 1961.[8]

USS LANGLEY

Even the USS LANGLEY was outfitted with a pigeon loft. Located on the fantail, it included nesting, training, and trapping areas for about one hundred birds.[9] When a pigeon returned to the ship, a bell would ring to let the pigeon master know there were birds to recover.[10]

One aviator noted that "The attempt to train pigeons to return to a ship was a great failure, but provided an excellent supply of squab for the mess."[11] Eventually, the pigeons were removed and the space converted into the executive officer's cabin.

PIGEON ADVENTURES IN PENSACOLA

Pigeons figured prominently in several adventures of Naval aviators in the Pensacola area. A 1919 survey of all facilities aboard Naval Air Station

Pensacola identified Building No. 215 as a wooden pigeon house, 40 feet by 15 feet that was built sometime between April and May 1918.[12]

A worker at the Naval Air Station recalled the pigeon loft:

Now I don't remember just when but we had a pigeon loft at the Naval Air Station, it came to life some time during WWI, it was located just inside the northeast corner of the high brick wall and they had good-size cages for the pigeons to exercise (sic) in. And if you are going to have pigeons you naturally will have young pigeons or squabs. In order to raise pigeons you have to get rid of the young. I never did know

Figure 2. To the uninitiated, one would think a pilot merely flipped a pigeon over the aeroplane's side. In actuality, the pilot forcefully threw the pigeon well clear of the aeroplane to avoid deadly whirling pusher propellers, murderous razor thin rigging wires, and lethal tail surfaces. (Courtesy U.S. Navy)

just what happened to the excess pigeons but if I was in charge I would soon find out what to do with them. I knew the man very well who was in charge of this job he was a friend of mine, the man in charge was Frank Lee, he was a United States chief petty officer. He lived in Pensacola practically all his life and he died here about a year or so ago. Every knew him and everybody like him and they called him Red Lee.

The 8 (sic)_16s and the F5L planes would go on navigation hops and when they went they would take cages with these homing pigeons in them and of course each of these homing pigeons would be marked with bands on their legs and they would at sea let the pigeons go and the pigeons would fly back to this fellow Red Lee and when they got back there he would make records of them and put them back in the big cage.[13]

An article in the December 1922 *Air Station News* noted:

A large number of homing pigeons are bred and trained on this station for reliable communication service. These pigeons are taken up in the planes and released while in the air at points distant from the station. Fastened to their legs are small receptacles, in which can be placed messages written on thin paper. When released from the planes they at once fly back to this station where the pigeon trainers are on the watch for them to receive the messages they carry.[14]

At least 90 trained pigeons were at the Naval Air Station in 1924, used "When the large seaplanes take long flights along the coast or into the gulf."[15]

GOD BLESS PIGEON NUMBER 19-P.F.-1

As noted in a previous chapter, four U.S. Navy balloonists set a distance record when they launched from Pensacola and traveled more than 700 miles before landing at Murdock, Illinois.[16] Homing pigeons were used during this flight to update the Naval Air Station on the balloonist's progress. One message from pigeon "N.A.S.—19-P.F.-1,"[b] launched by Lieutenant Theodore

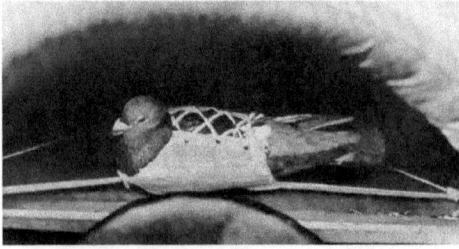

Figure 3. When carried aboard aeroplanes pigeons were sometimes placed in a holder to keep their wings from flapping. (Courtesy of author)

C. Lonnquest who was in charge of the flight read:

No. 2   Day 24 . Balloon 5601. 6: 45a.m.

Now in Jasper county, Mississippi, heading 178 degrees.   About 150 miles from Pensacola. Wind, 18mph. Have 18 bags ballast. Alt. 2,600. Everything O.K. but have been fired at several times. LONNQUEST[17]

OLD CORRY FIELD DEDICATION

Naval Air Station Pensacola's homing pigeons were used symbolically during the dedication of Old Corry Field on December 7, 1922: "...one hundred homing pigeons were released. These flying back to the Air Station visualized the connection of the Air Station with the Landing Field."[18]

PIGEON RECORDS

One Pensacola pigeon nearly broke a U.S. Navy record:

...a homing pigeon belonging to the loft at the Pensacola Naval Air Station has just completed a flight of 470 miles from Lake Okeechobee, Fla., to Pensacola. Released at daylight Sunday morning, the bird arrived at the air station yesterday afternoon at 2.

It is estimated that 20 hours was spent on the wing, for the bird would not fly at night and would probably stop for food. This bird is one of the younger ones at the Navy loft, being only 12 months old. The longest flight that it had previously made was from Biloxi, Miss. To Pensacola, a distance of 100 miles.

This flight sets a record for distance at the local pigeon loft and nearly equals the Navy record, which is 506 miles.[19]

PIGEON STATION DECOMMISSIONED

In July 1929, *The Pensacola Journal* sadly reported that:

Present indications point to the passing of one of the most interesting and fascinating branches of aviation. During the war and at other times in the past the use of pigeons as a means of communication was of great value and interest to naval aeronautics. Of late years their value has materially decreased due to the development and improvements in aircraft radio.

There is doubt in the minds of many naval officers whether the present value of pigeons as a means of communication warrants the amount of time and money that is being

---

[b] Each Naval air station pigeon loft in the United States had a specific designation letter to facilitate identification of stray birds. NAS Pensacola was "P.F."

spent on them. Because of this doubt, the navy department is investigating the subject. A questionnaire has been sent to all naval air stations and aviation units that still use pigeons asking for facts and opinions regarding the home flying birds.[20]

In September 1929, *The Pensacola Journal* noted that:

In accordance with recommendations from the Bureau of Aeronautics all pigeon activities at air stations other than lighter-than-air stations at Lakehurst and Guam, will be discontinued Oct. 1, 1929. All pigeon loft material allotted for that purpose will be either surveyed or turned over to other activities. Pigeons will be concentrated at Anacostia to be sold to the highest bidder.[21]

Figure 4. Identified as "Peerless Pilot," this is one of the homing pigeons that operated out of Naval Air Station, Pavillac, France during World War I. He delivered 196 messages. (Courtesy U.S. Navy)

In October 1929, *The Pensacola Journal* related that:

The navy department will discard the homing pigeon and abandon their training and upkeep, now that the radio and the airplane may be invoked to do the work the faithful bird has so long performed with accuracy and devotion. Thus does a machine age rudely shatter the cherished institutions handed down from a more romantic past. The pigeon which so long filled the role of man's faithful messenger in times of crisis and desperate need will, like Othello, find his occupation gone. The radio has come to carry the same word which the swift-flying bird once carried and perhaps with more speed and better assurance of getting to the desired destination. The radio will not only carry the message with greater celerity but will also return the answer long before the devoted pigeon could have gotten well started on its perilous journey. Pigeons, it is explained, will still be used and employed in isolated sections where the radio may not reach. Thus when machines fail animals must serve. The horse pulls out the mired motor, and it was not the radio but the dove that brought the twig to old Noah when he was floundering among the reefs around Mt. Ararat. – Selma Times-Journal.[22]

In a final article entitled "Bye Bye, Pigeons," the Air Station News wrote:

Our pigeons, which have been enjoying their beautiful new loft for nearly a year will soon be shipped to Anacostia to be sold to the highest bidder, as the Bureau of Aeronautics has decided that the only air-craft to rate pigeons in the future (that is, after 1 October, 1929) will be the lighter-than-air craft at Lakehurst and Guam.[23]

Pigeons had flown the Pensacola coop for the last time.

NOTES

[1] *Flying Officers of the United States Navy 1917-1919.* Atglen, PA: Schiffer Publishing Ltd, page 72.

[2] Howeth, L.S. (1963). *History of Communications-Electronics in the United States Navy.* [Electronic Version]. Retrieved from http://earlyradiohistory.us/1963hw01.htm#1footnote.

[3] Pride, A.M. (1998). Sea Legs. In Wooldridge, E.T. (Ed.) *The Golden Age Remembered.* Annapolis, MD: Naval Institute Press, page 53.

[4] Sudsbury, E. (1967). *Jackrabbits to Jets.* San Diego, CA: Neynesch Printers, page 103.

[5] *Flying Officers of the United States Navy 1917-1919.* Atglen, PA: Schiffer Publishing Ltd, page 69.

[6] Van Wyen, A.O. (1969). *Naval Aviation in World War I.* Washington, DC: Chief of Naval Operations, page 30.

[7] Rausa, R. (1976, August). Turntables and Traps. *Naval Aviation News,* pages 9-19.

[8] Malin, C.A. (1971). *The United States Navy's World of Work: Nearly 200 Years of Evolution.* [Electronic Version]. Retrieved from http://www.history.navy.mil/faqs/faq78-1.htm.

[9] Tate, J.R. (1998). We Rode the Covered Wagon. In Wooldridge, E.T. (Ed.) The *Golden Age Remembered.* Annapolis, MD: Naval Institute Press, page 66.

[10] Rausa, R. (1976, August). Turntables and Traps. *Naval Aviation News,* pages 9-19.

[11] Tate, J.R. (1998). We Rode the Covered Wagon. In Wooldridge, E.T. (Ed.) *The Golden Age Remembered.* Annapolis, MD: Naval Institute Press, page 66.

[12] Bennett, F.M. (1919, March 15). *Historical Data, Collection of.* Special Collection, 73-6, Box 2, John C. Pace Library.

[13] Forster, A.E. (1974, December). *NAS Allegany.* Pensacola Historical Society.

[14] Pigeons. (1922, December 5). *Air Station News, Pensacola, Florida.*

[15] Raby, J.J. (1924, December 28). Naval Aviation Training. *The Pensacola Journal.*

[16] Navy Balloon Pilots Return. (1920, March 28). *The Pensacola Journal.*

[17] Navy Balloon Pilots Return. (1920, March 28). *The Pensacola Journal.*

[18] Dedication of Corry Field. (1922, December 20). *Air Station News, Pensacola, Florida.*

[19] Pigeon Arrives At Yard After 470-mile flight. (1925, April 28). *The Pensacola Journal.*

[20] Barber, B.B. (1929, July 23). Daily News Of The Air Station. *The Pensacola Journal.*

[21] Daily News Of The Air Station. (1929, September 10). *The Pensacola Journal.*

[22] The Carrier Pigeon To Go. (1929, October 10). *The Pensacola Journal.*

[23] Bye Bye, Pigeons. (1929, September 20). *Air Station News,* Pensacola, Florida.

CHAPTER 25
THE FUTURE

Figure 1. Draftsman's drawing of the 24-passenger Pensacola Metal Aircraft Corporation's all-metal flying boat to be built in Pensacola in 1930. (Courtesy *Pensacola News Journal*)

GREAT PROMISE

We end our exploration of Pensacola's early aviation history in 1929 if for no other reason than two decades seem to bring an attractive symmetry to this book. This does not imply that the future does not hold extraordinary achievements for Pensacola's aviators, for it is quite the contrary. Several remarkable events in the 1930s alone warrant special consideration. For example, the first flight of the Pensacola Metal Aircraft Corporation's all-metal flying boat is at hand; a million dollar, 37-ton, four-engine French flying boat named the *Lieutenant de Vaisseau Paris* will be destroyed in Pensacola Bay; the world famous aviatrix Amelia Earhart will visit; Pensacola's first regular airline service will be achieved, although it will vanish fairly quickly as commercial aviation attempts to gain a small foothold on the Gulf Coast; and Pensacola's trials and tribulations in trying to obtain airmail service will finally come to fruition.

However, other events will unfold that will rob Pensacola of her relative innocence and bring to a close a relatively carefree period in her aviation history. First, the aching effects of the Great Depression, precipitated by the stock market crash of 1929, will soon overwhelm Pensacola and thousands of her citizens will be out of work. Second, the U. S. government, in the shape of the Aeronautics Branch of the Department of Commerce, will begin to rein in the barnstormers, flying circuses, and itinerant pilots as they seek to bring order and safety to a chaotic and dangerous flight environment. Finally, the war clouds of World War II will begin to swirl and manifest themselves in an unimaginable expansion of Naval aviation training activities across Northwest Florida as the United States prepares to enter this great conflict.

Above all else, it is important to note that Pensacola's early aviation advocates had done their job well. Both military and commercial aviation was firmly established in Pensacola and the two would prosper side by side throughout the next several decades. As Lieutenant Barrett Studley observed,[a] aviation in Pensacola had successfully survived the experimental stage of flight, passed ably through the rapid aeronautical advances precipitated by World War I, and positioned herself well to accommodate the growth of civil aviation throughout the country.[1] Pensacola was indeed well established on the "aviation map."

Figure 2. The capsized wreckage of the French million-dollar seaplane *Lieutenant de Vaisseau Paris* is lifted from Pensacola Bay. (Courtesy Pensacola Historical Society)

FINAL VIGNETTE

I leave you with one final tribute to the pioneers of Pensacola who sacrificed their lives during this era in the development of aviation. Jack Lincke wrote it:[b]

When the sun is setting across the entrance to Pensacola bay, throwing its crimson guidons across the blue of the sky and water, when it is quiet on the beach, one can almost picture these squadrons of pioneers cruising in the eon depths above. They fly past in formations of antiquated sea and land planes, the old pushers, odd-shaped Burgess, the Wrights. Their muffled motors are barely audible. It would be difficult

---

[a] A former flight instructor at Pensacola in the 1920s, Lieutenant Studley wrote several books on U.S. Navy flight training.

[b] Newspaperman, author, scriptwriter, stunt double and civilian flyer, Linke earned his own wings of gold and designation as a Naval aviator at Pensacola in 1931.

to imagine these men deprived of the privilege of flying. Surely somewhere in the scheme of things it must be that their wings still beat the cool, clear air in the endless gulf of sky.

They pass slowly, winging their solitary way straight into the mellow glow of the receding sun, almost invisible specks against the twilight. A deeper note is carried on the evening breeze and behind these fly the squadrons of the later planes, the spirit of the men who have died recently in flight training operations. One can see the Vees of N.Y.s, the Corsairs, N9s and Hawks.

The formations pass from view into the fading twilight, winging their way into eternity on the solo flight from which there is no landing. Perhaps those who fly now, hope that if their number goes up, they will find a blank file waiting in one of the formations, a place among comrades in these honored squadrons of the pioneers.[2]

NOTES

[1] Studley, B. (1929). *Practical Flight Training.* New York: The Macmillan Company, page 3.

[2] Lincke, J. (1931, May 19). Development of Aviation Claimed Many Lives. *The Pensacola Journal.*

SOURCES

There was certainly nothing subtle about my research. Once I realized that there was very little material on Pensacola's early aviation history, I simply sat down in the microfilm room of the University of West Florida's John C. Pace Library and proceeded to read every available copy of *The Pensacola*

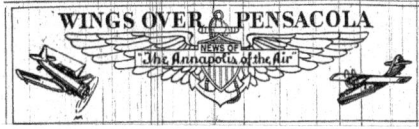

Courtesy *Pensacola News Journal*

Journal from 1908 to 1938. Remarkably, there are relatively few missing issues, and the staff works miracles to keep their ancient microfilm readers and printers in good working condition. The *Pensacola Evening News* is also available and covers the period January 1906 to December 1912. The West Florida Regional Library on West Gregory Street also maintains microfilm holdings of *The Pensacola Journal* that are duplicative of the University of West Florida's holdings, although not quite as extensive.

At the Naval Air Station Pensacola library microfilm and hard copies of the *Air Station News, Pensacola, Florida* (now known as the *Gosport)* are available from its inception in 1921. While focusing entirely on news of the air station, the introduction of landplanes is captured as is the training pipeline and curricula of this era.

However, despite my fascination in observing early twentieth century history unfold in nearly real time, I must confess that reading microfilm is still tedious and unforgiving to any inattentiveness. As the hours grew long and my eyes tired in a reading session, I am certain that I missed some significant events in Pensacola's aviation history.

Surprisingly, there are few extant photographs of civilian aeroplanes and aviators of this era. While *The Pensacola Journal* published hundreds of photographs of aviation achievements outside Pensacola, and the U.S. Navy documented its own deeds, newspaper photographs of local civilian aviators and their early flying fields are rare. This puzzles me as *The Pensacola Journal* covered in detail just about every local military or civilian aviation achievement, incident or visit. I am convinced photographs exist and that they are just held in private hands.

Another disappointment was the lack of coverage of Pensacola's aviation contact with Mobile and New Orleans in their newspapers. While *The Pensacola Journal* trumpeted various "first flights" from Pensacola to each city, a search of Mobile and New Orleans newspapers to verify details revealed nothing. I am at a loss to explain this phenomena as both cities were exceptionally aviation minded and were certain to send reporters to cover such significant events.

Pensacola is blessed with several outstanding repositories of local history. The University of West Florida's Special Collections Department, National Museum of Naval Aviation, Naval Air Station Pensacola Library, West Florida Historic Preservation Inc, the West Florida Regional Library, Pensacola Historical Society, and Historic Pensacola all contain a myriad of personal papers, photographs and memorabilia. While there is some duplication of holdings across all these wonderful organizations, their strength lies in their dedicated archivists who are more than willing to provide detailed assistance to broad searches.

For those interested in additional readings on early aviation in Pensacola, I would like to provide a small, select bibliography. These publications are readily available in local libraries or online bookstores. I consider Mr. Pearce's book to be the finest text ever written on the history of early Naval aviation in Pensacola.

Brown, W.J. (1994). *Florida's Aviation History*. Largo, FL: Aero-Medical Consultants. An outstanding resource for Florida aviation history.

Delaney, M. M. (Ed.). (1989). *The Cradle*. Pensacola, FL: Pensacola Engraving Company. Published to commemorate Naval Air Station Pensacola's 75th anniversary, this extremely well written text is chockablock with station photographs.

Deacon, S.M. (Ed.) (1975). *From Pirates to Pilots, A Pictorial History of Pensacola Navy 1528 to Present*. (2nd ed.). Pensacola, FL: Pensacola Engraving Co. Another excellent source of information and photographs of early Naval aviation at Pensacola.

Lazarus, W. C. (1951). *Wings in the Sun*. Orlando, FL: Tyn Cobb's Florida Press. The original reference source on Florida aviation history.

McGovern, J.R. (1976). *The Emergence of a City in the Modern South: Pensacola 1900-1945*. No publisher listed. While little is written about aviation, this book presents an outstanding study of Pensacola's culture and economic development.

Pearce, G.F. (1980). *The U.S. Navy in Pensacola from Sailing Ships to Naval Aviation (1825-1930)*. Pensacola, FL: University Presses of Florida. My personal goal when I began this project was not to cite Mr. Pearce as a reference on every fact.

INDEX